Röss · Learning and Teaching Mathematics using Simulations

How to use this book and the simulations

All simulations can be accessed at http://mathesim.degruyter.de/jws_en/

More information about this book can be found at
http://www.degruyter.de/cont/fb/ma/detail.cfm?id=IS-9783110250053-1

The electronic version – the e-book in PDF format – is linked to numerous single files. It is possible that your system will display the documents in PDF A mode in which these links are deactivated. In this case you can activate the links by making the following changes to the settings in Adobe Reader or Adobe Acrobat Professional with the file opened:

Menu Edit/Preferences/Documents/View Documents in PDF/A Mode/Never

The e-book gives you access to the download package for unpacking and installation to your local system. It also contains a PDF file giving links to the unpacked files on your system.

Dieter Röss

Learning and Teaching Mathematics using Simulations

Plus 2000 Examples from Physics

De Gruyter

Mathematics Subject Classification 2010: Primary: 97M20; Secondary: 97M50.

Prof. Dr. Dieter Röß
Fasanenweg 4
63768 Hösbach-Feldkahl
Germany
E-mail: dieter.roess@t-online.de

ISBN 978-3-11-025005-3
e-ISBN 978-3-11-025007-7

Library of Congress Cataloging-in-Publication Data

Röss, Dieter, 1932.
    Learning and Teaching Mathematics using Simulations: plus 2000 examples from physics / by Dieter Röss.
        p. cm. – (De Gruyter textbook)
    ISBN 978-3-11-025005-3 (alk. paper)
    1. Mathematics – Study and teaching – Simulation methods. 2. Physics – Study and teaching – Simulation methods. 3. Mathematics – Textbooks. 4. Physics – Textbooks. I. Title.
    QA20.C65R6713 2011
    510.71 – dc22
                                                                    2011007424

Bibliographic information published by the Deutsche Nationalbibliothek

The Deutsche Nationalbibliothek lists this publication in the Deutsche Nationalbibliografie; detailed bibliographic data are available in the Internet at http://dnb.d-nb.de.

Setting: Da-TeX Gerd Blumenstein, Leipzig, www.da-tex.de
Printing and binding: Hubert & Co. GmbH & Co. KG, Göttingen
♾ Printed on acid-free paper

Printed in Germany

www.degruyter.com

Preface

The idea of writing this *digital book* was born during discussions among a circle of friends[1] of the following questions: Author

(a) Why are physics and mathematics so unpopular at School?
(b) Why are there not more school leavers that are eager to study natural sciences and technology?
(c) Why do the large majority of first year students dismiss the very good subject-related and professional career opportunities in these professions?

Already, in the final years of school, mathematics and physics are considered to be *hard subjects*. Universities grudgingly accept that the mathematical knowledge of many school leavers is insufficient for taking up subject studies and needs to be improved by bridging courses.

A shockingly large number of students already fail in the first semesters of university. This will have serious consequences for the future welfare of our society, as we urgently require a sufficient number of well qualified young professionals in scientific and technical jobs to succeed the current generation of scientists and engineers.

It is easy to understand why the younger generation choose those *soft subjects* at university for which they feel better equipped and where they see better chances of success. The fact that the monetary concerns of finding a job later are not considered to be crucial in subject choice can actually be considered as a likeable attitude in students.

Why is it that mathematics and physics are considered to be so *difficult*? In fact, these should benefit from not being *rote learning subjects*: if one has understood a specific physics or mathematics problem within its context, one can forget the small details, since they can be reconstructed from the larger context.

It is obvious that, in our schools, one often does not manage to achieve this state of understanding and insight into the mathematical structure and laws of nature; the instruction therefore cannot provide the wonderful experience of having *understood* something. Thus, physics can indeed become a *cumbersome* subject full of incomprehensible and disconnected formulas and tedious calculations, and mathematics an art of *computation* that is build on memorization, and which increases in complexity from

---

1 Leading members of the *Deutschen Physikalischen Gesellschaft* (DPG; with 57 000 members, the largest society of its kind world wide), of the *Wilhelm and Else Heraeus Foundation* (WEH-Stiftung), and individual colleagues from the physics community, among them in particular Prof. Dr. Siegfried Grossmann and Prof. Dr. Werner Martienssen.

the times tables up to integration, while the fundamental ideas and deeper connections never become clear to the learner.

As the PISA study showed in 2003, this dilemma has developed in recent decades to such an extent that the level of mathematics and physics at German schools has declined from an earlier assumed top position to a "measured" weak, mediocre level. Similar declines in the standards of mathematics and physics have been reported in other countries.

What is the reason for this problem? We think that one of the most important reasons can be found in the **subject-specific education of future teachers at the universities!** Teacher training has often been treated as a stripped down appendix to the education of scientists. However, school teachers determine, in their respective subjects, the quality of education and the interests of the next generation! Their very important role in society as *multiplicators* has been neglected in relation to the education of future researchers representing a given subject. The resulting lack of recognition for students preparing to become teachers has certainly contributed to the lack of young teachers available to fill open positions in physics and mathematics.

Two developments in the immediate past have worsened this situation and made it clear that a turnaround is necessary:

- Educational policy has given, for good reason, more prominence to didactics and pedagogical studies, but has limited the duration of studies very rigidly. This, however, has meant that not much time is left for studying the subject matter.
- Students experienced the Bologna process in Europe as a transformation of the traditional freedom of academic study into stricter control, of a kind experienced at the schools that they had just fled. Overloaded syllabuses, and nearly continuous inspection of study progress with early crucial examinations led to early selection and a high failure rate.

  The attempt to serve "old wine in new skins", that is, to cram the traditional degree programs and the subject matter, which has grown due to scientific progress, into a shorter bachelor degree, has led to partial chaos and a general unhappiness with study conditions.

In 2005 *Siegfried Grossman* entered into discussions with the author with the conclusion that it is a fundamental mistake to mix the subject-specific education of teachers with that of researchers.[2] They demanded specially developed *sui generis* curricula for studies preparing for the teaching professions, which are directed at the future teaching job and that take into account the available time, which in Germany is limited by trainee teachers having to study two different subjects.

---

2 S. Grossmann, D. Röss: "Thesen zum Lehramtsstudium Physik – Plädoyer für eine eigenständige Lehrerausbildung", Physik Journal© 2005 4 (2005) Nr. 10, page 49

Understanding the subject and connections between different areas should be paramount, as opposed to detailed knowledge and specialized skills. In 2006, the DPG produced a careful analysis and documentation and thus made *sui generis* studies a general demand of colleagues involved in the DPG.

In order to realize this vision, it would be counterproductive to base our actions in relation to schools and students on past conditions or wishful thinking. We need to accept today's conditions, as well as the technical possibilities, in a positive spirit. The gymnasiums (German secondary schools that lead to the "Abitur", their final examination, allowing entrance to universities) should no longer be institutions for the elite, but should, in future, lead half of all children to the "Abitur". Access to a high level education could be similarly improved in many other educational systems around the world.

Our children and grandchildren are growing up in world with many stimulations and diversions, but have media skills that neither their parents nor grandparents had, for example their knowledge of, and playful dexterity with, the media and technological devices. This *digital book* is the attempt to put the abovementioned studies on a foundation that makes use of these skills and dexterity.

In this book, an important subset of the mathematical foundations is embedded in a systematically evolving text and presented with the help of numerical simulations, and is visualized in numerous ways. The PC takes care of the often tedious calculations. Thus, the user can concentrate on understanding the subject matter, the context and the algorithms used.

Since all the simulations are interactive and can, in many cases, also be used for scenarios that are totally different from those given, the students are thus given a quasi *"experimental"* access to mathematics. We make use of the fact that a visual impression is more intensive and permanent than a heard or read one, and that experience based on one's own actions results in deeper understanding than the mere reception of someone else's knowledge. In addition, playfulness is given free range to visually experience and grasp the intellectual stimulation and aesthetic beauty of mathematical structures.

The book provides colleagues in physics and mathematics with a thesaurus of simulations for the development of their own curricula. In addition to textbooks, this thesaurus gives physics students the possibility of a deepened understanding of fundamental mathematical notions and physical phenomena. Future teachers can, during their own training, experience the potential of modern media for the realization of interactive lessons in mathematics. Interested high school students can attempt a light-hearted introduction to a higher level of mathematics; they will probably have less trouble with the techniques used than some older people.

For the simulations, the package *Easy Java Simulation* (EJS) is used, which provides a simple fast-tracked introduction to the development of simulations in *Java*. The files produced with EJS are very transparent, and can be easily changed and reused as building blocks for one's own developments. The author considers EJS to be a prime candidate to become the standard program for didactically oriented simulations.

The authors of *EJS*, *Francesco Esquembre* and *Wolfgang Christian*, have allowed me to supplement this text, which is primarily an introduction to mathematics, with more than 2000 physics-based simulations, for which I owe them many thanks. *Francesco Esquember* has also assisted me personally in numerous ways with the creation of the mathematical simulations. I also thank *Eugene Butikov* for allowing me to include his wonderful cosmological simulations.

I want to give many thanks to *Siegfried Grossmann* for the dedication and care that he has applied to critically reading the text and simulations, and for the many valuable hints, which have contributed to the final version. *Ernst Dreisigacker*, the general manager of WEH-foundation, has supported me with the careful correction of details and with lively discussions.

Over the last three years I have had many involved discussions with *Werner Martienssen* about a book with a similar goal, i.e. to assist in reforming and improving the physics education of future teachers, and which is due to be published soon.[3] The idea to write this *digital introduction to mathematics* came up during these discussions.

The staff members at De Gruyters have done a great job in the production of this complicated publication. My special thanks go to Dr. von Friedeburg and to Ulrike Swientek for their personal engagement and permanent encouragement. Katherine Thomasset was a great partner in the final translation of the German original into English.

I want to thank my wife Doris for the loving understanding with which she tolerated my absentmindedness while this work was written. I promise improvement!

16 May 2011   Dieter Röss

---

3 "Physik im 21. Jahrhundert: Essays zum Stand der Physik". Editors Werner Martienssen and Dieter Röss, Springer Berlin 2011

Contents

Guide to simulation technique

Please use the deeply structured table of contents for the mathematical text, for the e-book in addition the search function of the Acrobat Reader.

The following index is intended as a door opener to the simulation technique used and to the mathematical simulations. In the e-book the specific pages can be directly addressed with links.

The number of a simulation in the index corresponds to its order of appearance in the text.

E – Physics-Simulations

F – Developing simulations with EJS

G – EJS

1 Introduction

1.1 Goal and structure of the digital book

This book is available in two versions: a printed one and an electronic one. All users can access the simulations at http://mathesim.degruyter.de/jws_en/. If you have chosen the eBook, or are accessing it through your library, the PDF is linked to the simulation files via the internet. The eBook also gives you, via the "supplemental material" button, access to download the complete software package (> 700 MB) for offline-use; the PDF in that package is linked to all simulations via hyperlinks. Buyers of the print version who would like to download and install the software to their local system can obtain access by contacting info@degruyter.com for registration. The directories of Figures 1.1 and 1.2 correspond to the download solution. It illustrates selected mathematical methods that are important for the presentation and the understanding of the contexts of physics.

The foundations of these mathematical methods are introduced by example. The programmability and computational power of the PC is used to visualize these methods, undertake calculations, change parameters, present connections interactively and present computation processes via interactive simulation and animation, in an interesting manner. In addition, playfulness is given free range. This presentation is also intended to make the beauty and aesthetics of mathematics visible.

The material provided in this book allows the user to penetrate mathematical structures and tools in an **experimental** manner. In particular, topics have been chosen that are difficult to imagine in an abstract manner, such as complex numbers, infinite sequences, transitions to the limit, fields, solutions of differential equations and so on.

All individual simulations contain extensive descriptions and suggestions for experiments. The user can always interactively engage with the simulations, and in many cases pre-programmed functions can be edited or new ones can be introduced. After some initial training in the *EJS* (*Easy Java Simulation*) program the user can open all files, change them and develop them further.

With one exception, Java programs have been used that were either created from OSP scratch or taken from the freely available internet projects *Open Source Physics* (*OSP*) and *EJS*. EJS

Our own mathematical simulations were created with the *EJS*-program that has been developed by *Francisco Esquembre*. Due to its graphical user interface, this program immensely simplifies the development and modification of simulations in comparison to "classical programming" in Java. This program and its documentation are contained in this book, but are also freely available on the internet.

We will, however, abstain from explaining the mathematical and computational techniques in systematic detail. The in-depth study of the mathematical and numerical methods will be left to specialist textbooks[4].

The pictures contained in the the pdf file mostly show screen shots of the respective simulation. When in the caption of such a picture *simulation* is clicked at for the first time a small context menu appears that asks, as a security measure against viruses, whether this file should be opened. You may confirm this and mark a check box, to avoid this dialog in future. The simulation will then be started immediately after clicking on simulation.

Where the reader might wish to learn more about the topic, *links* to internet pages have been inserted next to the text. They often point to *Wikipedia*-pages, from where further navigation is easy. These links are resolved in the outlined text boxes on the margin. All simulations can be accessed individually at http://mathesim.degruyter.de/jws_en/. Buyers of the printed book can also ask for download of the complete file package with interlinked text at info@degruyter.com for operation at their PC.

The appendix in Chapter 11 contains a short introduction to the EJS program and an extensive collection of simulations from all areas of physics, which have mostly been created using this tool. In order for these simulations to run on your computer, the *Java Runtime Environment* (*JRE*) must be installed, which you can download for free from the *SUN*-homepage in the latest version, using the link given at the border.

It is advisable to follow the suggestion to install the JRE into \Programs\JAVA\. For newer EJS simulations with 3*D-Rendering* you can download the *Java 3D* program from the same page.

1.2 Directories

The book consists of 3 units:

- The continuous **interlinked book text** *e-Exmat* as a *PDF* file of around 30 MB.
- A **secondary directory** *workspace* containing a directory tree, which is ordered according to topics and authors. It contains more than 2000 simulation files of around 650 MB size, of which around 1000 are executable *jar* files[5] that can be activated from inside the text. The *launcher* files among them also link to many

---

4 For example: *Mathematischer Einführungskurs für die Physik* 9. Auflage, Siegfried Grossmann (Teubner 2008) ISBN 3-519-33074-1; *Open Source Physics – A User's Guide With Examples* Wolfgang Christian (Pearson 2006) ISBN-10: 080537759X and ISBN-13: 9780805377590; *Mathematische Grundlagen für das Lehramtsstudium Physik*, Franz Embacher (Vieweg + Teubner 2008), ISBN 978-3-8348-0619-2.

5 *jar* files can be executed by themselves.

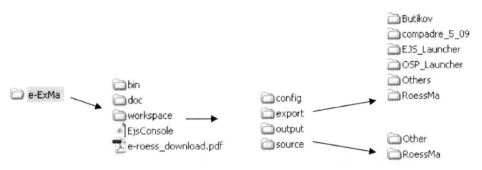

Figure 1.1. Main directory tree.

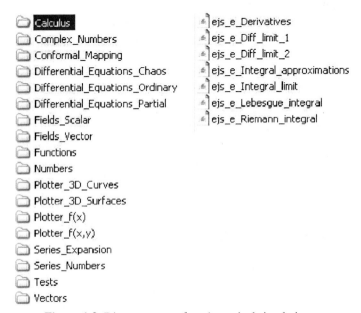

Figure 1.2. Directory tree of mathematical simulations.

secondary files for the individual simulations. The remainder are saved as *xml* files.[6]
- The **EJS console** for opening the *xml* files and working on the *jar* files, supplemented by documentation on the EJS program. The console does not have to be installed. It is contained in the program and can be started directly.

---

6 *xml* is an abbreviation for *Extensible Markup Language*. For our purposes xml files are text files that contain the instruction for the simulations. They cannot be executed by themselves, but are opened from the EJS console, from where the corresponding jar file is easily created via assembling the required Java library components. To view or directly change the *xml* file one can, for example, open it with *notepad*. An introduction to *EJS* is given in the appendix.

The main directory **e-ExMa**, which can be copied to an arbitrary place on the hard disk, contains the text file **e-Roess_download.pdf** the **Ejs console** and the directory **workspace**, with all simulation files. In **doc** are some documentation files for EJS and in **bin**, some configuration files and library files of the console. *Workspace* contains as sub-directories **export** for all executable jar files and **source** for the xml files that are meant to be opened from the console. In **output**, html files are saved while working with the console.

Export is divided into **RoessMa** for the mathematical simulations of the ongoing text, **Butikov** and **compadre** for the physics simulation files of the appendix. **Other** in the directory source contains further physics simulations in *xml*- and *jar*-format.

It is advisable to create links for the pdf text file and the *console* on the screen (in Windows: *Desktop*), in order to find these files quickly without searching for them. Please take care not to change the deeper directory structures, otherwise some hyperlinks will not work.

As long as the simulations are accessed from the text, you do not have worry about the directory structure, since it is saved in the hyperlinks. However, as soon as you want to edit a simulation from the console, you are asked for the location of the file.

The directories **RoessMa** in **export** and **source** are structured into directories in the same way according to topics. This is illustrated in Figure 1.2 for the example of the sub-directory **Calculus** with 6 individual simulations. The initially empty directory **Tests** is intended to save the data for your own experiments. This setup prevents the original files from being overwritten by mistake.

1.3 Usage and technical conventions

Most simulations are interactive. The user has several alternative ways of intervening, although not necessarily in parallel.

Individual points or elements of the graphical presentation can be "pulled" with the *mouse* and thus parameters can be changed. In this case, the mouse pointer changes into a hand symbol when it is positioned on the element.

Numerical values of different parameters can be entered into *number fields*. However, this change only becomes active if the enter key has been pressed and the text field, which turned yellow when entering text, becomes clear again. If the text field turns red, a mistake with the input has occurred (often a comma has been used instead of a full stop as decimal point; the correct format is, for example, 12.3 instead of 12,3).

From a *list of options*, given functions or parameter values can be selected with the mouse.

With *sliding controls*, individual parameters can be changed continuously or in steps.

Functions that are displayed in a *text field* can be changed or rewritten from scratch. Again, the changed function is submitted with enter.

When formulas are written in printed text, we often use short hand notation, by unspoken convention:

- for items that are ambiguous and that can be misinterpreted as text, like ab for a times b or $\sin a$ for $\sin(a)$;
- for items that can be misunderstood by software as formatting characters for text, such as x^2 for $x*x$ or $x\char`\^2$;
- for those special characters that cannot be interpreted by programs, such as $\dot{y}$ for $\frac{dy}{dt}$ for derivatives with respect to time.

For the input for numerical programs such as *EXCEL/VBA*, *Java*, *VBA* or *Mathematics* the notation must be unambiguous.

The fundamental rule is: all parts of the formula must be entered directly via the keyboard without the use of special characters. Combined characters must be mapped to an equivalent number of keyboard characters, in order for these to be correctly identified by the program. (Example: y' as a derivative combined from two keyboard characters; a unique text like "derivative with respect to t" might also be interpreted by a program). In particular, the following notations have to be noted:

- Addition and subtraction: $a + b, a - b$;
- Multiplication: $a*b$;
- Do not omit brackets: $a*\sin(b)$;
- Division: a/b; $(a + b)/(c + d)$;
- Power: $a\char`\^b$;
- Exponential function: $\exp(a)$;

Many simulations use a *parser* to translate the formulas entered as text into *Java* Parser format. In this case, the following notation is permissible, which can also be used recursively.

$atanh(x)$	$ceil(x)$	$cos(x)$	$cosh(x)$	$exp(x)$	$frac(x)$
$floor(x)$	$int(x)$	$ln(x)$	$log(x)$	$random(x)$	$round(x)$
$abs(x)$	$acos(x)$	$acosh(x)$	$asin(x)$	$asinh(x)$	$atanh(x)$
$sign(x)$	$sin(x)$	$sinh(x)$	$sqr(x)$	$sqrt(x)$	$step(x)$
$tan(x)$	$tanh(x)$	$atan2(x, y)$	$max(x, y)$	$min(x, y)$	$mod(x, y)$

Here we have $acos = $ *arc cosine*, $cosh = $ *hyperbolic cosine*.

The important expression $atan2(x, y)$ prevents the ambiguities of the *arctan* by automatically yielding the correct angle in the second and third quadrant; here x and y are the sides of the triangle involved and x is opposite to the angle.

$step(x)$ is a very interesting function in practice. It switches at $x = 0$ from 0 to
1. If one wants to superimpose the function $f(x)$ to the function $g(x)$ from $x = x_1$
onwards, then this can be written as $g(x) + f(x)*step(x - x_1)$. For some simulations
the *Math* package is used together with Java for calculations. In this case, the functions
are prepended with *Math* as follows: $Math.cos(x)$.

Further details about the functions and terminology used in Java can be found from
many sources on the internet, for example via searching for *Java & Math*. Just try the
link at the border.

1.4   Example of a simulation: *The Möbius band*

As an example of the possibilities of interactive simulations as they will be used in the
following chapters, Figure 1.3a shows a rotating *Möbius band* in three dimensional
projection. Among the closed bands in space the Möbius band is characterized by the
fact that it makes half a twist, and thus, during one circulation, both sides are covered
by a traveller; it has "only one surface". In the picture of the simulation, one sees the
formulas for the three spatial coordinates with the variables p and q, which contain
two parameters a and b, which can be changed with sliders. The slider for a changes
the number of half twists, while the other one changes the height of the band. If a
non-integer number is chosen for the number of half twists a, the band can be cut,
and rejoined with another number. If this number is even, one obtains normal bands
with 2 surfaces. It this number is odd, one obtains a Möbius band that has additional
twists.

The formulas for the three space coordinates, as well as the time dependent animation component, can be edited, i.e. they can be changed. Using the same simulation,
arbitrary animated surfaces in space can be visualized. The ability to edit opens a wide
training field for the advanced understanding of functions that describe three and four
dimensional processes. Figure 1.3b shows two examples from the simulation of Figure 1.3a. On the left a simple band with a full twist, and on the right a Möbius band
with one and a half twists, were calculated.

The text pages of the simulation contain extensive descriptions, hints for many
alternatives of the 3D-projection, and suggestions for experiments. Figure 1.4 shows
the description window that appears next to the simulation when it is opened. In this
example, it contains 4 pages:

Introduction with a description of the simulation and its controls;
Visualization with hints about the possibilities of $3D$ projection;
Functions for discussions of the mathematical formalism;
Experiments with suggestions for experiments that make sense.

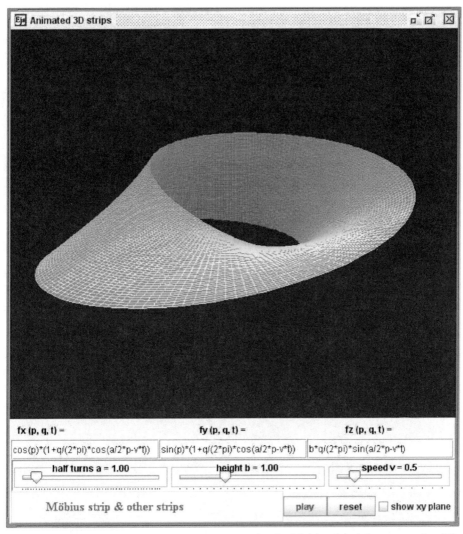

Figure 1.3a. Simulation. The figure shows a simple Möbius band in perspective 3D-projection. The three function boxes contain the parameter representations for the space coordinates. The variables p and q vary in the range $-\pi$ to π. The parameter a that determines the number of twists (here $1/2$) can be changed with a slider, as well as the parameter b that controls the height of the band. The z-component can be periodically modulated for $v > 0$ with the angular velocity v (play button) in order to create the impression of a rotating band. Using a check box, the xy-symmetry plane can be shown or suppressed.

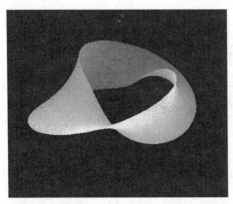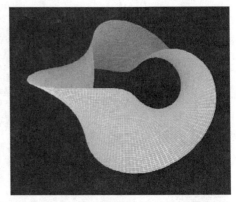

Figure 1.3b. Examples from the simulation in Figure 1.3a; on the left a simple loop with $a = 1$, on the right, a Möbius band with an additional twist ($a = 1.5$).

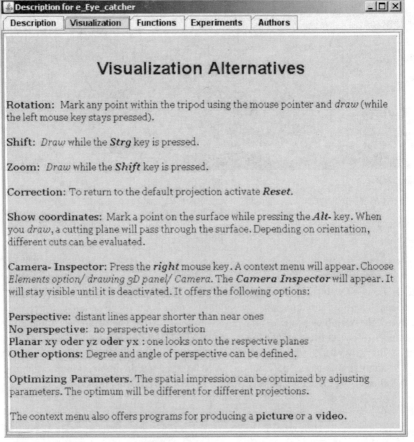

Figure 1.4. Description window of the simulation. Here it contains 4 pages, of which the visualization page is opened. Please test the possibilities after opening the simulation!

In the figure, the page for **Visualization** is opened. It describes easy possibilities for different three dimensional presentations:

- Rotation;
- Translation;
- Zoom;
- With or without perspective distortion;
- Projections along one of the three axes.

You are encouraged to use this example to try the different means of experimentation before you start with the next chapter.

2 Physics and mathematics

2.1 Mathematics as the "Language of physics"

Physics (Greek φυσική, "the natural") researches the fundamental interactions in nature. Already, the natural philosophers of antiquity thought at a deep level about the phenomena in the cosmos as well as in nature as it surrounds us, and their methodology was mostly of a *qualitative*, descriptive and often speculative nature.

The great progress of physics in modern times is due to capturing the natural phenomena by measuring them *quantitatively*, and comparing the results of measurements with assumed relationships (hypotheses). This process allows, via the interaction of experiments and hypotheses, the evolutionary development of original hypotheses to physical "theories" that are applicable in ever larger generality.

Thus theories are well-tested hypotheses for relationships in nature, which are formulated in the language of mathematics. It is an initially startling result of the interaction over hundreds of years between experimental and theoretical physics, that a plethora of individual phenomena can be described in terms of standard theories, whose mathematical formulation only require a few symbols or lines of symbols. We list here the *Schrödinger equation* as the fundamental equation of quantum mechanics, *Maxwell's equations* of electrodynamics and the *Navier–Stokes equation* of hydrodynamics.

These fundamental equations can become numerically difficult or sometimes even close to unsolvable when applied to specific cases within the huge variety of phenomena embraced by the original theoretical model.

However, a large number of phenomena of practical importance can be described by very simple mathematical models, which are also easily applied to individual cases. These include nearly all those phenomena that are important for engineering and its effect on our daily life.

Using a suitable level of abstraction of the theories, one can model an ever larger variety of phenomena in a single theory – it is for a reason that the *world formula*, from which all theories of physics can be derived, is the ever desired, but not attained goal of theoreticians.

It is an unanswered question of epistemology whether "the book of the universe is written in the language of mathematics", as expressed by Galileo Galilei[7]; that is, whether physical theories describe the reality of nature, or whether, as formulated by the *positivists* among the natural scientists, as long as the model has never been

---

7 *Il Saggiatore* 1623, paragraph 25

falsified by detrimental experimental evidence. The first school of thought includes the natural philosophers of antiquity as well as Einstein and Schrödinger, among the modern scientists; the second school of thought is characterized by names such as Born, Bohr and Heisenberg.

In any case, mathematics provides physics with a powerful tool.[8] On the one hand, `Math` the corresponding methods were developed directly when studying physical questions, as was the case with calculus, developed by *Isaac Newton* (1663–1727) and *Gottfried Wilhelm Leibniz* (1664–1716) while studying the movements of planets. On the other hand physics, when studying new questions, sometimes makes use of methods that have been developed in the frame of pure mathematical and logical reflections, such as in the case of the general theory of relativity, which made use of the non-Euclidean geometry developed by *Georg Friedrich Bernhard Riemann* (1826–1866).

The strict formulation of mathematical relationships in the highly specialized language of mathematics appeals to the expert with its convincing stringency, transparency and terseness. To the beginner, however, this kind of presentation may seem confusing and too complex. In this text we will choose, as far as possible, a concrete description, and otherwise refer to specialized textbooks and internet links.

2.2 Physics and calculus

The *State* of nature at a given point in time could be fixed via a photographic snapshot `Infinity` and described with words. In a mathematical and physical picture, this would correspond to a description of nature via formulas in which time does not appear. Thus, already, many states, for example equilibria, can be described via simple mathematical equations.

In addition, physics examines and describes *changes* in nature,[9] and, as a rule, these changes happen as functions of time. This enables theories to describe the development of a current state from its conditions at an earlier point in time. More important is the ability to predict a *future* state from the knowledge of the current state; this ability empowers the techniques based on it to achieve a desired, *future* effect.

For the deeper understanding and practical application of physics, the knowledge of differential calculus is necessary, since the *changes* (derivatives) and the sum of their effects (integrals) have to be considered. Without this understanding, physics becomes a collection of more or less disconnected formulas, which are only applicable to very limited cases. Thus the calculation of results may become a nuisance for school students, and this blocks their insight into the simplicity and beauty of the relationships between mathematics, physics and technology.

---

8 Immanuel Kant V, 14 (Akademie-Verlag edition) says: *in every kind of philosophy of nature, only so much science can be found, as there is mathematics to be found in it.*
9 Immanuel Kant AA XXII, 134 (Akademie-Verlag edition): *Physics is the science of moving forces, that are inherently connected to matter.*

But the mathematical operations and methods that are needed for a basic under-standing of the subject are not really difficult. Using suitable visualizations, the no-tions used can be easily grasped. Using a computer for calculation and for creating the visualization media (diagrams, animations, simulations) it becomes easy to to put this into practice.

3 Numbers

We first want to remind the reader of the different kinds of numbers that are used in arithmetic and to visualize their relationships with the elementary arithmetic operations. Here the number is the *operand* on which a certain *operation* is applied (arithmetic operations such as $+, -, *, /, \hat{}, =, >, <$, and logical operations such as and, or, not, if-then, otherwise, ...).

The definitions of numbers and arithmetic operations are, up to the complex numbers, synchronized in such a way that for any number z, the following fundamental rules of arithmetic operations apply. Here ( ) means that the operation in brackets is executed first.

$$z_1 + z_2 = z_2 + z_1 \qquad \text{(Commutative law of addition)}$$
$$z_1 \cdot z_2 = z_2 \cdot z_1 \qquad \text{(Commutative law of multiplication)}$$
$$(z_1 + z_2) + z_3 = z_1 + (z_2 + z_3) \qquad \text{(Associative law of addition)}$$
$$(z_1 \cdot z_2) \cdot z_3 = z_1 \cdot (z_2 \cdot z_3) \qquad \text{(Associative law of multiplication)}$$
$$(z_1 + z_2) \cdot z_3 = z_1 z_3 + z_2 z_3 \qquad \text{(Distributive Law of multiplication)}$$

Consequence: It does not matter in which sequence the operations are executed.

Shorthand notation in the text: $z_1 z_2 \equiv z_1 \cdot z_2$; $z^2 = zz$; $z^3 = z^2 z (= z\hat{}3)$; ....

The following sections will introduce different families of numbers in consecution of their historical development. The requirement to apply certain operations, which had been introduced for a certain kind of number, without restrictions, led to successive extensions of the usual concepts of numbers.

3.1 Natural numbers

The natural numbers are $1, 2, 3, 4, 5, \ldots$ in the set of natural numbers, which in mathematics is referred to as $\mathbb{N}$.[10] Additions can be executed without limit as well as multiplications, which are to be understood as multiple additions: $3 \cdot 4 = 4 + 4 + 4$.

$\boxed{\text{Numbers}}$

In using number notation, one differentiates between *ordinal numbers* (*the third* – in an imagined sequence) and *cardinal numbers* (*three pieces*). Toddlers of 3–4 years often know the ordinal numbers up to 10 and they can also execute simple additions via counting. The more abstract notion of the cardinal number children mostly understand only when they start school; in addition, even for the adult, the number of units that can

---

10 Historically, special symbols have been introduced for the number sets (see link in the margin).

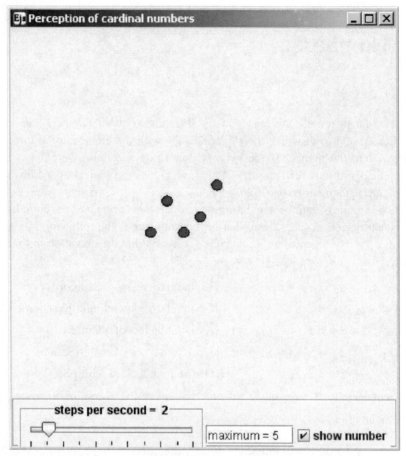

Figure 3.1. Simulation. Spontaneous grasping of the number of elements in a set (cardinal numbers) A random number generator produces red points, whose number lies between 1 and the maximum number in the number field (in the figure the maximum number is 5, 5 are shown). The sets change with a frequency that can be adjusted with the slider from 1 to 10 per second.

be grasped at a glance is quite limited (to around 5–7, which is also what intelligent animals are capable of); for fast calculations with cardinal numbers, the relationship is memorized or simplified in our thoughts ($5+7 = 5+5+2 = 10+2 = 12$). If one realizes this fact, one gains a deeper understanding of the difficulty that children have with learning the elementary rules of arithmetic. Simply assuming the memorized routines, which are present in an educated adult, leads to severely underestimating the natural hurdles of understanding that the children have to overcome when they learn arithmetic.

The simulation in Figure 3.1 visualizes the sharp threshold that nature imposes for spontaneously grasping the number of elements of a set. In this simulation, points are shown in a random arrangement that can be spontaneously grasped as a group. The number changes with a frequency that can be specified between 1 and a maximum number. You can establish experimentally where your own grasping threshold lies. The description pages of the simulation contain further details and hints for experiments.

Even numbers are a multiple of the number 2; a *prime number* cannot be decomposed into a product of natural numbers, excluding 1.

The lower limit of the natural numbers is the unity 1. This number had a close to mystical meaning for number theoreticians of antiquity, as the symbol for the unity of the computable and the cosmos. It also has a special meaning in modern arithmetic as that number which, when multiplied with another number, produces the same number again.

There is, however, no upper limit of the natural numbers: for each number there exists an even larger number. As a token for this boundlessness, the notion of infinity developed, with the symbol ∞, which does not represent a number in the usual sense.

Already, the preplatonic natural philosophers (*Plato* himself lived from 427–347 BC) worked on the question of the infinite divisibility of matter (If one divides a sand grain infinitely often, is it then still sand?) and time (if one adds to a given time interval infinitely often half of itself, will that take infinitely long?)

Zenon of Elea (490–430 BC) showed in his astute paradoxes, *Achilles and the tortoise* and *the arrows*,[11] that the ideas of movement and number theory at the time were in contradiction to each other.

Subtraction is the logical inversion of addition: for natural numbers it is only permissible if the number to subtract is smaller than the original number by at least 1.

Division is the natural inversion of multiplication. For natural numbers it is permissible if the dividend is an integer multiple of the divisor – 6 : 2 = 3.

3.2 Whole numbers

In order for the operation of subtraction to be always possible, we have the extend the natural numbers by zero (the "neutral" element of addition) and the negative numbers to the set $\mathbb{Z}$ of *whole* numbers.

The introduction of the zero as a number was, historically, not a trivial step. Zero is connected with the notion of *nothing*, and for the pre-socratic natural philosophers

---

11 *Achilles and the tortoise*: during a footrace, Achilles allows the tortoise a head start. When he has reached its starting point, the tortoise has already crawled a certain distance further. When he reaches this point, the tortoise has again had a head start, and so on. Thus Achilles cannot reach the tortoise. (*Solution: Convergence of the geometric series, which had not been recognized at the time*).

The flying arrow: at every instant, the arrow is situated at a certain point, which is fixed in space; therefore, the arrow is at rest in every instant. Therefore the arrow cannot move. (*Solution: In every instant, the arrow does not have only a location but also a velocity; this differential concept was not recognized.*)

it was a fundamental question, whether *nothing* (the emptiness, something that is not) can exist or not.

Parmenides of Elea (around 600 BC) taught that *nothing* cannot exist, but that every space has to filled by something, which leads to the paradoxical logical consequence that movement is impossible and everything is unchangeable. The atomicists *Leukipp* (5th century BC) and *Democrit* (460–371 BC) taught, however, that the world consists mostly of nothing (today we would call it a vacuum), in which objects that consist of material atoms can move.

In the 3rd century BC, the zero was imported from the east in connection with the campaigns of Alexander the Great as a sign indicating the position in the decimal number system (as today in 10,100). The concept of zero as a whole number was only reached in the 17th century.

The set of whole numbers contains the natural numbers as a subset

$$\text{Whole numbers:} \ \ldots, -3, -2, -1, 0, 1, 2, 3, \ldots.$$

Multiplication is admissible without exception if one defines: $(-1) \cdot 1 = -1$; $(-1) \cdot (-1) = 1$ and $(0) \cdot (1) = 0$.

In the domain of whole numbers, the symbol for positive infinity must be necessarily supplemented by negative infinity $-\infty$; both numbers are not numbers in the usual sense.

Division can be applied for whole numbers, as for natural numbers, if the divisor is contained as a factor in the dividend, i.e. if the division works out, as for $-30 : 5 = -6$.

Division by zero is not a well defined inversion of multiplication:

for integers a, b, c

$$\frac{b}{a} = c \quad \text{uniquely leads to } a = \frac{b}{c}$$

$$\text{for } \frac{0}{a} = 0 \quad a \text{ can be any number.}$$

and is therefore excluded. The expressions $0 \cdot \infty$, $\frac{\infty}{\infty}$ and $\frac{0}{0}$ are not defined.

Whole numbers are visualized as a discrete ladder on the number line (see Figure 3.2). Arithmetic operations amount to jumping back and forth on this ladder – in the same way that toddlers indeed make calculations with natural numbers by counting.

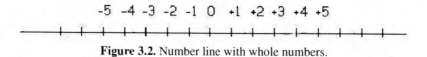

Figure 3.2. Number line with whole numbers.

3.3 Rational numbers

In order for division to be admissible in all cases, except for division by *zero*, the whole numbers have to supplemented by the "broken numbers" to form the set of *rational* numbers $\mathbb{Q}$. Rational numbers contain the whole numbers as a subset:

$$\text{rational number} = \text{whole number} : \text{whole number} = \frac{\text{whole number}}{\text{whole number}}$$

Examples: -5; $-\dfrac{3}{2}$; $1175/1176$; 3; $1{,}1357$; $5{,}28666666\ldots$; $\ldots$

When written as a decimal number, rational numbers are decimal numbers with a remainder that has a finite length or with periodically repeating digits.

There is no largest rational number.

It is clear that whole numbers are rare, special cases of rational numbers. Between two subsequent whole numbers, there are infinitely many rational numbers.

When dealing with the set of rational numbers, division by zero is still not a well defined inversion of multiplication and remains formally excluded. If one starts from the concept of zero as the limit of a sequence of nearly infinitely small positive or negative rational numbers, then division by zero would be equivalent to the definition of nearly infinite positive or negative numbers. In this symbolic sense, division by zero can be associated with a sequence that has a limit of $\pm\infty$.

Taking the power of a number is defined for rational numbers as repeated multiplication, with the whole numbers n as exponent, by:

$$A^n = A \cdot A \cdot A \cdot A \qquad n \text{ times}$$

$$A^0 = 1; \qquad A^{-n} = \frac{1}{A^n}.$$

Taking the nth root is the logical inversion of taking the power. In the domain of rational numbers, root-taking is possible:

- if the exponent n of the root is odd;
- *or* when for even root exponents the original number (the *radicand*, the number under the root sign) is positive;
- *and* if in both cases the operation results in a rational number, which is only the case for rare radicands, that can be reduced to fractions of powers, for example $\sqrt{6.25} = \sqrt{\frac{625}{100}} = \frac{25}{10} = 2.5$.

3.4 Irrational numbers

If an operation applied to a rational number (e.g. root-taking, the limit of an infinite sequence of rational numbers) leads to a number that is not a rational number, i.e. if it cannot be written as a ratio of 2 integers that is representable as a finite or periodic decimal fraction, then this number is defined to be an *irrational* number. Here the term *irrational* is given for historical reasons, as a demarcation from the *rational* (numbers that are ratios) and has no secondary meaning of irrational = unreasonable or unthinkable.

If one applies the operations mentioned above to irrational numbers, then this does not lead to a more generalized number.

Rational numbers constitute a countable set – they can be ordered in such a way that they constitute a countable sequence. The irrational numbers, on the other hand, do not constitute a countable set. In this sense, there are more irrational than rational numbers.

3.4.1 Algebraic numbers

The need to introduce numbers that are not rational was recognized by the Pythagoreans (*Pythagoras*, 570–510 BC, mathematician and natural philosopher in the Greek colony Metapont in southern Italy) during their reflections on the calculation of right triangles with a hypotenuse c and legs a and b.

In the domain of integers, there are only a few solutions to a right triangle, the Pythagorean triples, which are often used in homework problems: $(3, 4, 5; 6, 8, 10; 5, 12, 13; 8, 15, 17; 7, 24, 25; 9, 12, 15; 10, 24, 26;$ etc.)

Theorem of Pythagoras: $$a^2 + b^2 = c^2 \rightarrow c = \sqrt{a^2 + b^2}$$

Example of an integer solution: $$c = \sqrt{3^2 + 4^2} = \sqrt{25} = 5$$

Example of a rational solution: $$c = \sqrt{\left(\frac{3}{2}\right)^2 + 2^2} = \sqrt{\frac{25}{4}} = \frac{5}{2}$$

Example of an irrational solution: $$c = \sqrt{1^2 + 1^2} = \sqrt{2}$$

Numbers that are generally obtained as the solutions of polynomial equations with rational coefficients, i.e. that are their roots, are designated as *algebraic* numbers. They include both rational and irrational numbers.

3.4.2 Transcendental numbers

Irrational numbers that are not a root of a polynomial with rational coefficients are called *transcendental* numbers.

Here *transcendental* simply means *going beyond* the rational numbers and does not have any *mystical connotation* whatsoever.

The most common transcendental numbers are the *circle number* π and the *Euler number e* (written in blocks of five in the following)

$$\pi = 3.14159\,26535\,89793\,23846\,26433\,83279\,50288\,41971\,69399\,37510\ldots$$

$$e = 2.71828\,18284\,59045\,23536\,02874\,71352\,66249\,77572\,47093\,69995\ldots.$$

It is a characteristic feature of transcendental numbers that they are *limits* of infinitely often repeated operations (additions, multiplications, formations of continued fractions, root taking, etc.; see below).

3.4.3 π and the quadrature of the circle, according to Archimedes

Using the example of the number π, it will be demonstrated how this transcendental number of high practical importance can be obtained as the limit of a sequence. We follow the famous train of thought originally devised by Archimedes.

Using the theorem of Pythagoras and the formula for the area of a triangles with baseline a and height h, i.e. $F = \frac{1}{2}ah$ the mathematicians and surveyors of Egypt and antiquity were able to reduce the area of an arbitrary surface that is bounded by straight lines to that of a square of the same area, whose length is given by a square root, i.e., generally an irrational number; even today the unit of surface area for arbitrarily bounded surfaces is still the "square meter".

The "quadrature of the circle", as a paradigm for calculating the area of a surface that is bounded by curved lines, however, remained unsolved for a long time.

The famous inventor and mathematician *Archimedes* (287–221 BC), who lived in the Greek colony Syracuse in Sicily, found a royal road to this end, which was only further developed nearly 2000 years later, and which represents the beginning of working with convergent, infinite sequences and with limits.

His method, which starts with a polygon that is inscribed or circumscribed to a circle (Figure 3.3), will be demonstrated in brief due to its historical significance. He uses the theorem of Pythagoras, the formula for the area of a right triangle, and symmetry considerations. From the above it follows that the baselines of the triangles constituting the polygons with n corners are given as a simple function of n when doubling n. The following diagrams visualize the procedure. The first regular polygon, a yellow square, is circumscribed around the circle filled in gray; a second colorless square is inscribed in the circle.

The inscribed polygon has a smaller area then the circumscribed one; The true value for the circle lies between the two. It is immediately evident that halving the angle of division to obtain an octagon, which is blue filled, will make the differences smaller, and that this goes on with further doubling of N (a polygon with 16 corners also is shown in red). The sketch shows the first steps of the calculation for inscribed polygons with 2^N corners, with $N > 2$.

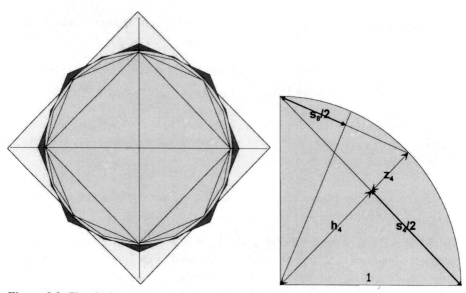

Figure 3.3. Simulation. Approximation of the circle via inscribed and circumscribed polygons. The simulation shows the approximations from the square to the polygon with 4096 corners.

The square, with which the calculation starts, consists of 4 equal right triangles, whose cathetuses for the unit circle under consideration have length 1. According to the theorem of Pythagoras, the hypotenuse of each triangle has the length $\sqrt{2}$. The height h_4 is obtained via the theorem of Pythagoras using $s_4/2$ and the hypotenuse 1 of the lower triangle. The distance z_4 is the difference between the radius 1 and the triangle height h_4. The transition to the octagon again proceeds with the theorem of Pythagoras via $s_4/2$ and z_4. As the following calculation shows, this algorithm can be repeated in the same way in factors of 2 towards a subdivision of the surface of the circle into ever smaller triangles. Thus this procedure results in *recursion formulas*, with which one can obtain the results of the Nth step from those of the $(N-1)$st step. We give the results for the *inscribed* polygon with n sides.

radius $r = 1$; index $n = 2^N$, with $N = 2, 3, 4, 5, \ldots$

$$s_4 = \sqrt{1+1} = \sqrt{2}; \quad h_4 = \sqrt{1 - \left(\frac{s_4}{2}\right)^2}; \quad z_4 = 1 - h_4 = 1 - \sqrt{1 - \left(\frac{s_4}{2}\right)^2}$$

$$s_8 = \sqrt{\left(\frac{s_4}{2}\right)^2 + z_4^2} = \sqrt{2}\sqrt{1 - \sqrt{1 - \left(\frac{s_4}{2}\right)^2}};$$

$$h_8 = \sqrt{1 - \left(\frac{s_8}{2}\right)^2} = \frac{1}{\sqrt{2}}\sqrt{1 + \sqrt{1 - \left(\frac{s_4}{2}\right)^2}}$$

recursive formula

$$s_N^i = \sqrt{2}\sqrt{1 - \sqrt{1 - \left(\frac{s_{N-1}^i}{2}\right)^2}};$$

$$h_N^i = \sqrt{1 - \left(\frac{s_N^i}{2}\right)^2} = \frac{1}{\sqrt{2}}\sqrt{1 + \sqrt{1 - \left(\frac{s_{N-1}^i}{2}\right)^2}}$$

circumference U, surface F: $\quad U_N^i = n s_N = 2^N s_N = 2^N \sqrt{2}\sqrt{1 - \sqrt{1 - \left(\frac{s_{N-1}}{2}\right)^2}}$

$$F_n^i = n\frac{s_N h_N}{2} = 2^N \frac{s_N h_N}{2} = \frac{2^N}{2}\sqrt{2 - \left(\frac{s_{N-1}}{2}\right)^2}.$$

In the following, the equations have been written out starting from the inscribed square $n = 4$ to the polygon with $n = 64$ corners. One realizes the iterated characters of the repeated root-taking of the side length $\sqrt{2}$ of the triangles making up the inscribed square.

$$s_4 = \sqrt{2} \qquad\qquad\qquad h_4 = \frac{1}{2}\sqrt{2}$$

$$F_4 = \frac{4}{4}2 = 2{,}0000$$

$$s_8 = \sqrt{2 - \sqrt{2}} \qquad\qquad\qquad h_8 = \frac{1}{2}\sqrt{2 + \sqrt{2}}$$

$$F_8 = \frac{8}{4}\sqrt{2} = 2{,}8284$$

$$s_{16} = \sqrt{2 - \sqrt{2 + \sqrt{2}}} \qquad\qquad h_{16} = \frac{1}{2}\sqrt{2 + \sqrt{2 + \sqrt{2}}}$$

$$F_{16} = \frac{16}{4}\sqrt{2 - \sqrt{2}} = 3{,}0614$$

$$s_{32} = \sqrt{2 - \sqrt{2 + \sqrt{2 + \sqrt{2}}}} \qquad\qquad h_{32} = \frac{1}{2}\sqrt{2 + \sqrt{2 + \sqrt{2 + \sqrt{2}}}}$$

$$F_{32} = \frac{32}{4}\sqrt{2 - \sqrt{2 + \sqrt{2}}} = 3{,}1214$$

$$s_{64} = \sqrt{2 - \sqrt{2 + \sqrt{2 + \sqrt{2 + \sqrt{2}}}}} \qquad h_{64} = \frac{1}{2}\sqrt{2 + \sqrt{2 + \sqrt{2 + \sqrt{2 + \sqrt{2}}}}}$$

$$F_{64} = \frac{64}{4}\sqrt{2 - \sqrt{2 + \sqrt{2 + \sqrt{2}}}} = 3{,}1365.$$

These formulas are fascinating in their aesthetic symmetry!

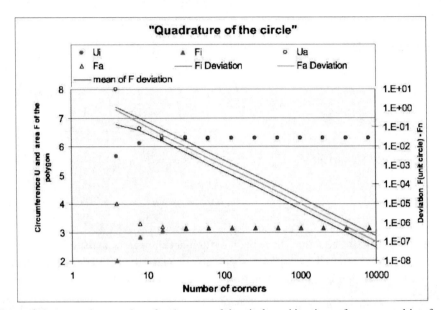

Figure 3.4. Approximate values for the area of the circle and its circumference resulting from the "quadrature of the circle" according to Archimedes. The abscissa shows the number of corners of the inscribed or circumscribed polygons. The circular points in the upper region of the ordinate are the approximations for the circumference, those in the lower region are those for the surface of the unit circle. The lines in the logarithmic scale show the difference in red for the approximation from the inside, green from the outside and blue for the average.

With a simple spreadsheet, this calculation that was rather tedious for Archimedes can be done quite quickly up to a high number of corners.

Then one sees how quickly the surface areas of the inscribed and circumscribed polygon approximate the number π (3.14159...) and the corresponding circumference approximates 2π. In Figure 3.4 they are shown for the square up to the polygon with 8192 corners (corresponding to $N = 2$ to $N = 13$). In addition, the respective differences of the surface area from π are given (logarithmic right-hand scale). Already for the 10th approximation (polygon with 1024 corners) the difference is only 10^{-5}.

Archimedes himself started with a *hexagon* and took the calculation up to a polygon with 92 corners and obtained his value for the *circular number* of 3.141635 (the symbol π for this number was only introduced in the 18th century); we suggest that you retrace the calculation of Archimedes.

3.5 Real numbers

Rational and irrational numbers together constitute the set of *real numbers* $\mathbb{R}$. They fill the number line *densely* (every arbitrarily small neighborhood of a real number on the number line contains at least another real number).

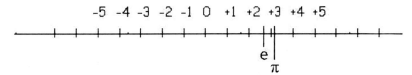

Figure 3.5. Number line with two transcendental irrational numbers.

Taking the power and root-taking with rational exponents is possible in the domain of real numbers, if the root exponent is odd ($\sqrt[3]{-1} = -1$; $\sqrt[3]{1} = 1$) or, for even exponents, if the argument of the root is positive via the following definitions:

$$\text{rational number } q = \frac{n}{m}; \; n, m \text{ integer}$$

$$\rightarrow \text{ for real numbers } x: x^q = x^{n/m} = \sqrt[m]{x^n} = (\sqrt[m]{x})^n.$$

The real numbers constitute the largest possible *ordered* set of numbers. For two real numbers a and b, it is clear whether a is larger than, equal to, or smaller than b:

$$a > b \quad \text{or} \quad a = b \quad \text{or} \quad a < b.$$

For applications in physics, the distinction between rational, irrational and transcendental numbers plays an important role, as their symbols express relationships in a formula that has been derived via a model. If the number π appears, circular symmetry or periodicity plays a role, while the appearance of e points to a problem involving growth or damping.

As soon as computations take place, irrational numbers are always approximated with finite accuracy via rational numbers. The formally excluded division by zero in the domain of real numbers loses its exceptional position, since it will always be the division by a very small, but finite real number.

The arithmetic operations can be interpreted as transformations or mappings on the number line. Addition and subtraction are translations where all numbers are shifted by the absolute value of the summand. Multiplication and division by n lead to stretching or compression of the number line by a factor n.

Division by $a > 1$ corresponds to a transformation of the range of numbers outside of the dividend to the range of numbers between the dividend and zero.

For the example $1/a$, $a = 1$ is mapped to itself, numbers $a > 1$ are mapped to the range 0 to 1, the closer to zero, the larger a is. Numbers $0 < a < 1$ are mapped to the domain larger than 1 and the further away from 1, the closer a is to zero.

3.6 Complex numbers

3.6.1 Representation as a pair of real numbers

The even numbered root of a negative radicand cannot be represented in the domain of real numbers, since for all real numbers x we always have $x^2 > 0$. For a polynomial of second degree with real numbers x, the generally known solution only yields numbers when the radicand is larger or equal to zero:

$$ax^2 + bx + c = 0 \text{ has the two solutions } x_{1,2} = \frac{-b \pm \sqrt[2]{b^2 - 4ac}}{2a}$$

$b^2 \geq 4ac \rightarrow$ real number as solution

$b^2 < 4ac \rightarrow$ no solution in the domain of the real numbers

in the simplest case $x^2 + c = 0$ we would have $x = \sqrt{-c}$;

then there exists for positive c, i.e. $c > 0$,

no solution in the domain of the real numbers.

To allow a solution for all c, including the positive ones, one extends the one-dimensional space to two-dimensional *number pairs* of real numbers, that are called *complex numbers* $\mathbb{C}$, and for which a *special multiplication rule* is agreed.

Complex numbers were first used in the 16th century in connection with roots of negative radicands by the mathematicians *Girolamo Gardano* and *Raffaele Bombelli*.

Complex numbers satisfy the following rules:

General definition of the complex number z as an ordered pair of numbers

$z = (a, b)$ a, b real numbers

addition rule $\qquad z_1 + z_2 = (a_1, b_1) + (a_2, b_2) = (a_1 + a_2, b_1 + b_2)$

multiplication rule $\quad z_1 \cdot z_2 = (a_1, b_1) \cdot (a_2, b_2) = (a_1 a_2 - b_1 b_2, a_1 b_2 + a_2 b_1)$

conjugate complex number definition: $\overline{z} = (a, -b)$;

this leads to $z\overline{z} = (a, b) \cdot (a, -b) = (a^2 + b^2, 0) \equiv a^2 + b^2$

division: $\quad \dfrac{z_1}{z_2} = \dfrac{(a_1, b_1)}{(a_2, b_2)} = \dfrac{(a_1, b_1) \cdot (a_2, -b_2)}{(a_2, b_2) \cdot (a_2, -b_2)}$

$$= \frac{(a_1 a_2 + b_1 b_2, -a_1 b_2 + a_2 b_1)}{a_2^2 + b_2^2} = \frac{z_1 \overline{z_2}}{z_2 \overline{z_2}}.$$

The main innovation relative to the "one-dimensional" real numbers is the multiplication rule. For numbers whose second component vanishes, the familiar multiplication rule for real numbers results; when their first components are zero, the product of 2 complex numbers is:

$$(a_1, 0)(a_2, 0) = a_1 a_2 = \text{sign}(a_1)\,\text{sign}(a_2)|a_1|\,|a_2|;$$

with sign(a_1): algebraic sign of a_1

$|a_1|$: absolute value of a_1.

For the product of two numbers, whose first components vanish, one obtains from the definition

$$(0, b_1)(0, b_2) = -b_1 b_2 = -\text{sign}(b_1)\,\text{sign}(b_2)|b_1|\,|b_2|.$$

The product is in both cases a one-dimensional, real number. The second case is equal to the first case except for an additional signum.

The practical justification for these rules follows from their consequences, historically particularly from the fact that, in the domain of number pairs defined in this way, the taking of roots with rational exponents is possible without restriction. In the simplest example: one looks for the solution of $z^2 = -1$:

$z^2 = -1$; the approach: $z = (a, b)$ leads to

$z^2 = (a, b) \times (a, b) = (a^2 - b^2, 2ab) = -1 = (-1, 0)$

comparison of coefficients yields:

$a^2 - b^2 = -1$ und $2ab = 0$.

The second equation gives $a = 0$ oder $b = 0$

the latter possibility is excluded, since we must have $a^2 \neq -1$

therefore $a = 0$ and thus $b = \pm 1$

thus $z_1 = (0, 1); z_2 = (0, -1)$.

The real numbers are a one-dimensional subset of the two-dimensional complex numbers (a, b), i.e. those with $b = 0$; thus the real numbers are rare exceptions among the complex numbers. The complex numbers with $a = 0$, i.e. $(0, b)$ are called "imaginary numbers". Their square is negative: $(0, b)(0, b) = -|b|^2 < 0$.

3.6.2 Normal representation with the "imaginary unit i"

In the usual notation, the *normal representation* of complex numbers distinguishes between the two components instead of their sequence in brackets via a marker in front of the second component, for which, following *Leonhard Euler* (1707–1783), the letter i is used (in electrical engineering one uses instead the letter j to distinguish the marker from the current i). The plus sign indicates that both components belong together.

Unfortunately the term "imaginary" number has become common for the second component, which may create mystical ideas about its specific character, as something not as easy to understand as a real number. However, there is no class of "imaginary numbers"; both components of the pair that form a complex number are real. The notation $5i$ does not refer to a "multiplication of 5 with i", but means that the second component of the complex number is 5.

The normal representation $z = a + ib$ simplifies the calculations, since one can use in it the usual multiplication rules for real numbers, if one takes into account the convention $i^2 = -1$. Thus the following rules for the normal representation have to be interpreted accordingly.

Let us consider an example:

$$z_1 z_2 = (a_1 + i b_1)(a_2 + i b_2) = a_1 a_2 + i^2 b_1 b_2 + i(a_1 b_2 + a_2 b_1)$$
$$= a_1 a_2 - b_1 b_2 + i(a_1 b_2 + a_2 b_1)$$

complex number:	$z = (a, b)$
real number:	$a = (a, 0)$
definition imaginary number:	$ib = (0, b)$
definition:	real component$(z) = \mathrm{Re}(z) = a$
definition:	imaginary component$(z) = \mathrm{Im}(z) = b$
definition imaginary unit:	$(0, 1) = i$
definition:	$z = \mathrm{Re}(z) + i\,\mathrm{Im}(z) = a + ib$
definition conjugate complex number:	$\bar{z} = \mathrm{Re}(z) - i\,\mathrm{Im}(z) = a - ib$
consequence:	$z\bar{z} = a^2 + b^2$
definition absolute value:	$|z| = \sqrt{z\bar{z}} \geq 0$

computation rules in normal representation

$$z_1 + z_2 = a_1 + a_2 + i(b_1 + b_2)$$

$$z_1 z_2 = (a_1 + i b_1)(a_2 + i b_2) = (a_1 a_2 - b_1 b_2) + i(a_1 b_2 + a_2 b_1)$$

$$\frac{z_1}{z_2} = \frac{z_1}{z_2}\frac{\bar{z}_2}{\bar{z}_2} = \frac{a_1 a_2 + b_1 b_2}{a_2^2 + b_2^2} - i\frac{a_1 b_2 - a_2 b_1}{a_2^2 + b_2^2}$$

$$= \frac{a_1 a_2 + b_1 b_2}{z\bar{z}} - i\frac{a_1 b_2 - a_2 b_1}{z\bar{z}}$$

$$i^2 = ii = (0, 1) \cdot (0, 1) = (-1, 0) = -1$$

in this specific sense i is the square root of (-1).

Using the *normal representation* the solution of the square root problem becomes clearer.

c real number

$$z^2 = (a + ib)(a + ib) = a^2 - b^2 + i2ab = c$$

c real $\rightarrow 2ab = 0$

a product vanishes if and only if one of the factors vanishes.

Therefore: 1st solution $a = 0 \rightarrow -b^2 = c$

$$b = \pm\sqrt{-c} = \pm\sqrt{c}\sqrt{-1} = \pm ic \qquad z = 0 \pm i\sqrt{c}$$

for $c < 0$

or: 2nd solution $b = 0 \rightarrow a^2 = c$

$$a = \pm\sqrt{c} \qquad z = \pm\sqrt{c} + i \cdot 0$$

for $c \geq 0$.

In the set of complex numbers, the square root of a real number always has two solutions. They are either both purely real or imaginary, depending on the sign which the root has taken.

The general solution for a quadratic polynomial with real coefficients a, b and c, with which we started, now reads:

$$z_{1,2} = -\frac{b}{2a} \pm \frac{\sqrt{b^2 - 4ac}}{2a} = \begin{cases} -\frac{b}{2a} \pm \frac{\sqrt{|b^2 - 4ac|}}{2a} & \text{for } b^2 > 4ac \\ -\frac{b}{2a} \pm i\frac{\sqrt{|b^2 - 4ac|}}{2a} & \text{for } b^2 < 4ac. \end{cases}$$

If a, b and c are themselves complex, the general formula is still valid, but not the distinction between two cases since the order relations $>$ and $<$ are not applicable for complex numbers.

What is the situation in the complex number space for the cube/third root and, in general, for odd root exponents? In the space of real numbers, there is always one real negative solution ($\sqrt[3]{-c} = -\sqrt[3]{c}$) for negative radicands. In the space of complex numbers, however, we obtain the following:

$$z^3 = c; \quad c \text{ real}$$

$$(a + ib)(a + ib)(a + ib) = (a^2 - b^2 + i2ab)(a + ib)$$
$$= a^3 - 3ab^2 + i(3a^2b - b^3) = c$$

since c is real $\rightarrow b(3a^2 - b^2) = 0$

either $b = 0$ or $(3a^2 - b^2) = 0$

1st solution $b = 0 \rightarrow a^3 = c$

$$a = \sqrt[3]{c} \qquad z = a = \sqrt[3]{c}$$

this is the always existing real solution

2nd solution $3a^2 - b^2 = 0 \rightarrow b^2 = 3a^2$

$$a(a^2 - 3b^2) = c \rightarrow a(a^2 - 9a^2) = -8a^3 = c$$

$$a = \sqrt[3]{\frac{-c}{8}} = -\frac{1}{2}\sqrt[3]{c}$$

$$z_2 = -\frac{1}{2}\sqrt[3]{c} + i\frac{\sqrt{3}}{2}\sqrt[3]{c} = \sqrt[3]{c}\left(-\frac{1}{2} + i\frac{\sqrt{3}}{2}\right)$$

$$b^2 = \frac{3}{4}(\sqrt[3]{c})^2 \rightarrow b = \pm \frac{\sqrt{3}}{2}\sqrt[3]{c}$$

$$z_3 = -\frac{1}{2}\sqrt[3]{c} - i\frac{\sqrt{3}}{2}\sqrt[3]{c} = \sqrt[3]{c}\left(-\frac{1}{2} - i\frac{\sqrt{3}}{2}\right)$$

two conjugate complex solutions

$$\frac{\sqrt{3}}{2} = \sin 120° = -\sin 240° \quad z_1 = \sqrt[3]{c}\cos 0°$$

$$z_2 = \sqrt[3]{c}\left(\cos 120° + i\frac{\sqrt{3}}{2}\sin 120°\right)$$

$$-\frac{1}{2} = \cos 120° = \cos 240° \quad z_3 = \sqrt[3]{c}\left(\cos 240° + i\frac{\sqrt{3}}{2}\sin 240°\right) = \overline{z_2},$$

$$\text{since } \sin 240° = -\sin 120°.$$

Thus three roots z_1, z_2, z_3 of $z^3 = c$ exist, of which one is real and the other two are complex conjugates of each other.

3.6.3 Complex plane

The complex numbers are mapped for visualization purposes to points in a plane, where the abscissa corresponds to the real number line and the ordinate corresponds to the complex number line, and distances on both are measured using real numbers.

The simple cubic equation $z^3 = c$ has three solutions in the space of complex numbers, of which one is real, and two are complex. As the last representation for $c > 0$ shows (for $c > 0$ the points are mirrored on the imaginary axis), the roots are situated symmetrically on a circle with radius 1.

In the diagram the cube roots are indicated as squares and the two square roots as circles.

The general polynomial of nth degree has, in the space of complex numbers, n roots according to *Gauss' fundamental theorem of algebra*. Figure 3.6 shows this for the second and third root of 1.

Taking into account the rules for addition and multiplication, all usual arithmetic operations known for real numbers can also be applied to complex numbers.

The complex numbers densely cover the complex plane, as the real numbers cover the number line densely. Unlike the real numbers, the complex numbers are, however, *no ordered* set, since they each consist of two real numbers, and therefore the relation $z_1 > z_2$ is not defined in general. However, they can be ordered according to absolute values $|z_i|$, which are real numbers.

The use of complex numbers (*complex analysis*) has many advantages in physics and engineering.

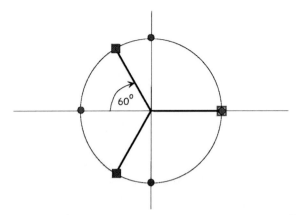

Figure 3.6. Roots in the complex plane: the blue circular points show $\pm\sqrt{1}$, the red circular points show $\pm\sqrt{-1}$ and the squares show $\sqrt[3]{1}$.

As shown for the example of the parabola, every algebraic equation has solutions in the domain of complex numbers (property of algebraic closure; *Gauss' fundamental theorem*).

In addition, every complex function that can be differentiated once can be differentiated an arbitrary number of times. Finally one can show with complex numbers relationships between individual functions that are independent in the domain of real numbers (e.g. exponential function and trigonometric function, see below).

3.6.4 Representation in polar coordinates

In the representation using polar coordinates, the absolute value $|z|$ gives the distance r from the origin and the ratio of the imaginary component to the real component is equal to the tangent of the angle ϕ to the real axis.

The following definition for the polar representation is applicable:

$$z = r(\cos\phi + i\sin\phi)$$

To obtain r and ϕ from z or vice versa the following relations apply:

$$r = |z| = +\sqrt{z\bar{z}} = +\sqrt{\mathrm{Re}^2(z) + \mathrm{Im}^2(z)}$$

$$\tan\phi = \frac{\mathrm{Im}(z)}{\mathrm{Re}(z)}.$$

Multiplication and division rules become:

$$z_1 z_2 = r_1 r_2[(\cos\phi_1\cos\phi_2 - \sin\phi_1\sin\phi_2) + i(\cos\phi_1\sin\phi_2 + \cos\phi_2\sin\phi_1)]$$

$$z_1 z_2 = r_1 r_2[\cos(\phi_1 + \phi_2) + i\sin(\phi_1 + \phi_2)]$$

$$\frac{z_1}{z_2} = \frac{r_1}{r_2}[\cos(\phi_1 - \phi_2) + i\sin(\phi_1 - \phi_2)].$$

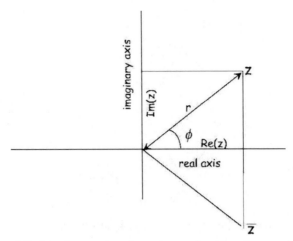

Figure 3.7. Complex numbers in representation via polar coordinates.

The polar representation shows a special position of the number zero in the domain of complex numbers, which is not visible in the domain of real numbers. It is the only complex number that does not have a direction associated with it, since $z = 0$ means $r = 0$, irrespective of the value of ϕ. This is also compatible with the tangent of ϕ being undetermined as a ratio of two zeros.

The number z corresponds in polar representation to the end point of a vector that starts at the origin with length r and makes an angle ϕ with the real axis.

3.6.5 Simulation of complex addition and subtraction

Pressing the ctrl key and clicking on the following pictures' *simulation* activates the interactive Java simulations, which demonstrate the complex operations *addition* and *subtraction*.

The operations are visualized in Figure 3.8 as mapping of a rectangular array of points in the z-plane to a u-plane shown on the right-hand side. The points are colour coded to show their assignment in the two planes. For the red point on the lower left corner of the arrays, the position vector is indicated. On the z-plane you can change the position of the red corner of the array as well as the tip of the green vector which is linked to it by pulling with the mouse. The u-plane shows the result of the complex operation. Clicking on the "initialize" button restores the original state. The distance between the points in the array can be adjusted with the slider. In particular, you can collapse the array to a single point.

In addition to the simulation, a text with several pages is shown. This text contains a detailed description of the simulation and hints for possible experiments.

The windows can be hidden or blown up to full screen size with the usual symbols on the top right; however, it makes more sense to blow up the simulation windows

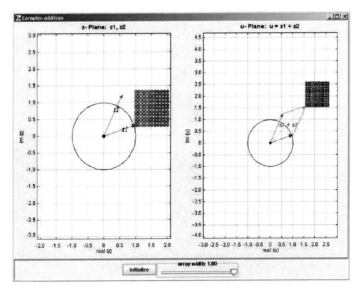

Figure 3.8. Simulation. Addition of a complex number z_2 to all numbers z_1 of a point grid. This grid is moved with the tip of the red arrow, which leads to the lower left corner of the array in the z-plane; in the same way $z_1 + z_2$ moves for all complex z_1 the whole complex plane. The supplementary sides of the parallelogram for the vector construction are drawn on the right-hand side.

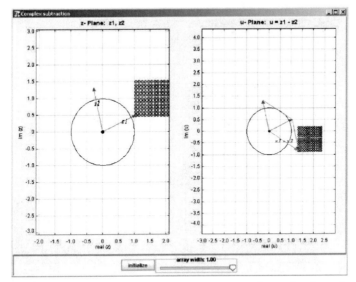

Figure 3.9. Simulation. Subtraction of a complex number z_2 from all points of a grid. In the left window the point array and the tip of the vector to be subtracted can be pulled with the mouse. The supplementary sides of the parallelogram of the vector construction are shown on the right-hand side.

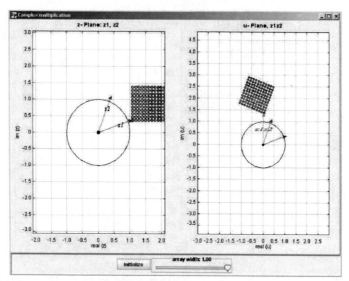

Figure 3.10. Simulation. The multiplication of z_1 with z_2 corresponds to a rotation of the vector z_1 by the angle of the vector z_2 in the mathematical positive sense (anticlockwise), while at the same time being stretched by the absolute value of z_2 (compressed, if the absolute value is smaller than 1). The point array and the complete plane is rotated via the angle of z_2 while being stretched by the absolute value of z_2.

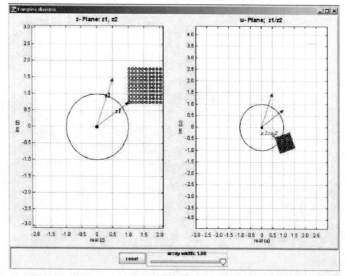

Figure 3.11. Simulation. Division corresponds to the rotation of z_1 in the mathematical negative sense (clockwise) by the angle of the vector z_2 while undergoing compression by its absolute value (stretching, if the absolute value is smaller than 1).

by pulling on one corner in order to preserve the quadratic structure of the system of coordinates. If you click on a point with the mouse, its coordinates appear in the window with a colored background. With the right mouse button you can access further options in the context menus.

The addition of two numbers corresponds in the complex plane to the addition of both vectors (according to absolute value and direction). The subtraction shown in Figure 3.9 corresponds to a subtraction of both vectors. Considered as a mapping of the z-plane, addition and subtraction are equivalent to a translation of the plane without rotation or change of scale.

3.6.6 Simulation of complex multiplication and division

The following two interactive pictures deal with the simulation of complex multiplication (Figure 3.10) and division (Figure 3.11). The presentation and handling is identical to that described above for addition and subtraction.

3.7 Extension of arithmetic

One can, of course, continue the extension of notion of a number from numbers to number pairs. The next step would be *quaternions*, which consist of four real numbers. Quaternions can be used for calculations in four dimensional spaces; for example relativistic physical systems with spin can be described using quaternions. We here refer to the subject literature.

Quater-
nion

When defining applicable rules for arithmetic operations, care is taken that the complex numbers constitute a subset. However, for these higher dimensional numbers, not all fundamental rules, which were given at the beginning of this chapter and remain valid up to the complex numbers, will necessarily hold, for example the rule of the commutativity of operations.

The *group theory* finally disassociates itself totally from the concept of the number, and defines arithmetic rules for elements, that can be numbers, but do not have to be numbers. A group (set) of elements is defined, and the rules for this group are defined in such a way that the application to elements of the group always yields a member of the group.

Group
theory

The rules applicable to groups are similar to those that we discussed at the beginning of our discussion of numbers. However, the earlier implicitly assumed role of unity (the neutral element) will be explicitly defined. For the example of the multiplicative composition we shall assume by definition:

1. The composition of two elements a, b of the group G is again an element of the same group (*closedness*) $a \times b = c \in G$.
2. The sequence of operations is unimportant as long as the order is preserved: $a \times (b \times c) = (a \times b) \times c$ (*associativity*).
3. There is a *neutral element* e in the group G, for which $a \times e = e \times a = a$.

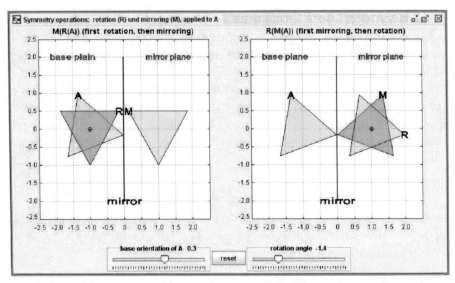

Figure 3.12. Simulation. Rotation and reflection of a pink equilateral triangle A. In the left window rotation (R) happens before mirroring (M); in the right window mirroring happens before rotation. The final result of the operations is colored in blue, the intermediate result in green. The initial orientation of A and the rotation angle applied can be chosen with sliders.

4. For every element a in G, an *inverse element* (mirror image) a^* exists, with the property to yield, when combined with a the neutral element: $a \times a^* = a^* \times a = e$.

A group is called *Abelian* or commutative if one is allowed to commute the operands: i.e. $a \times b = b \times a$ *(commutativity)*.

The set of integers $\mathbb{Z}$ with addition as the operation and zero as the neutral element is an Abelian group. Somewhat more complicated is the situation with the set of rational numbers $\mathbb{Q}$, multiplication as the operation, with 1 as the neutral element; here 0 would not have an inverse element.

The definition of group rules also makes it possible to have other objects than numbers as members of a group, as long as they satisfy the requested properties.

An example for such a group is the set of symmetry transformations *rotation, reflection* and *inversion*, through which a topological object such as a polygon is mapped to itself; compositions are then transformations that are applied consecutively. This example is a *non-Abelian* group; rotation before reflection yields a different result from reflection before rotation. This is visualized in the simulation shown in Figure 3.12. It shows the consecutive operations *rotation and reflection* and *reflection and rotation* applied to an equilateral triangle, whose initial orientation can be adjusted.

Group theory is one of the foundations of the arithmetic used in quantum theory; you are encouraged to study the discussion in the essay by *Schopper*.[12]

---

12 *Physik im 21. Jahrhundert: Essays zum Stand der Physik*, edited by Werner Martienssen and Dieter Röss, Springer Berlin 2011.

4 Sequences of numbers and series

4.1 Sequences and series

By repeated application of the same arithmetic operations on an initial number A, one creates a logically connected **sequence** of numbers, which show interesting properties (to guess the formation law of a sequence and thus to continue the initial numbers of a given sequence is a popular type of puzzle). Sequence

In the following the letters m, n, i, j are used to indicate the position of terms in sequences. They can be 0 or positive integers.

If there is no upper limit for the number of terms in a sequence or for the terms in a series ($m \rightarrow \infty$), we refer to an infinite sequence or series.

4.1.1 Sequence and series of the natural numbers

The particularly simple arithmetic sequence of the natural numbers is created via the repeated addition of the unit **1**; the individual term is characterized by the lower index $(1, 2, \dots)$, which itself is an increasing natural number.

$$A_1 = 1; \quad A_{n+1} = A_n + 1 \quad \text{for } n \geq 1 \rightarrow$$
$$A_n = 1, 2, 3, 4, 5, 6, \dots.$$

We now define the *difference quotient* for the terms of an arbitrary sequence with different indices i and j. This number is a measure for the change between two terms with different indices and thus for the growth of the sequence in the interval given by the indices:

$$\Delta A_{i,j} = A_i - A_j; \quad \Delta_{i,j} = i - j$$
$$\text{difference quotient: } \left(\frac{\Delta A_{i,j}}{\Delta_{i,j}} \right) = \frac{A_i - A_j}{i - j}.$$

For *consecutive* terms, the index interval is 1 and the difference quotient is equal to the difference between the terms:

$$\Delta A_{i,i-1} = A_i - A_{i-1}; \quad \Delta_{i,i-1} = i - (i - 1) = 1$$
$$\text{difference quotient: } \left(\frac{\Delta A_{i,i-1}}{\Delta_{i,i-1}} \right) = A_i - A_{i-1}.$$

For the sequence of the natural numbers, the difference between consecutive terms is constant and equal to 1. Therefore their difference quotient is also constant and equal to 1.

$$\Delta A_{i,i-1} = A_i - A_{i-1} = 1 \rightarrow \text{difference quotient} = 1.$$

The arithmetic sequence has constant growth of consecutive terms.

From the terms of a sequence one can, by repeated addition, obtain a logically connected series, whose partial sums are again terms of a sequence. For the above example this would be the *arithmetic series* S and the sequence of its partial sums S_n: [Series]

$$S = 1 + 2 + 3 + 4 + 5 + 6 + \cdots$$

$$S_1 = 1; \ S_2 = 3; \ S_3 = 6; \ S_4 = 10; \ S_5 = 15; \ S_6 = 21; \ \ldots$$

$$S_n = \sum_{m=1}^{n} A_m = A_1 + A_2 + A_3 + \cdots + A_n.$$

For the sum sign Σ (capital Greek letter *Sigma* (S)) the index m of the sequence terms A_m runs from the number on the bottom to the number on top.

For the arithmetic series, one can calculate the partial sums very easily from the indices. This rule is thought to have been discovered by Gauss when he was asked in school to sum up the numbers from 1 to 100. This rule is founded on the symmetry of the series: two numbers that are symmetrically positioned relative to the middle of the partial sum always add up to the same sum $(n + 1)$ and there are $n/2$ such pairs.

$$S_n = \frac{n}{2}(n + 1)$$

The sequence of the natural numbers does not have an upper limit. The sum over its subsets increases faster with increasing index, in quadratic dependence on the index:

$$n \gg 1 \rightarrow S_n \approx n^2/2.$$

4.1.2 Geometric series

As another example, we consider the sequence of powers of the real number a and the *geometric series* that is created from it via addition:

$$A_0 = 1; \ A_1 = a; \ A_2 = a^2; \ A_3 = a^3; \ A_4 = a^4; \ \ldots$$

$$A_n = a^n \quad \text{for } n \geq 0$$

definition: $\Delta A_{i,j} = A_i - A_j$; $\Delta_{i,j} = i - j$

$\Delta A_{i,j} = a^i - a^j$

$$\left(\frac{\Delta A}{\Delta}\right)_{i,j} = \frac{a^i - a^j}{i - j}; \quad \left(\frac{\Delta A}{\Delta}\right)_{i,i-1} = \frac{a^i - a^{(i-1)}}{1} = a^{(i-1)}(a - 1)$$

$$S_n = \sum_{m=0}^{n} a^m = 1 + a + a^2 + a^3 + \cdots + a^n.$$

For the special case of $a = 1$ the partial sums of the geometric series become an arithmetic sequence.

For a different from 1 the difference quotient depends on the index. For $a < 1$ it keeps getting smaller; the terms of the sequence decrease, and the partial sums increase ever slower. For $a > 1$ the difference quotient is positive and grows with the index; the terms of the sequence increase faster and faster, and the partial sums of the series even more so.

4.2 Limits

What happens if the index of the sequence or series becomes larger and larger, i.e. if it goes to infinity. Are the terms of the sequence getting larger and larger (in this case we call the sequence *divergent*), or do they approach a limiting value, i.e. the sequence is *convergent*? Does the value of the series grow to infinity or does it remain bounded, i.e. does it have a limit and is it convergent?

The sequence of the natural numbers obviously grows without limit as well as the value of the series; both are divergent:

$$\lim_{n\to\infty} A_n = \lim_{n\to\infty} n = \infty$$

$$\lim_{n\to\infty} S_n = \lim_{n\to\infty} \sum_{m=1}^{n} m = \infty.$$

What about the geometric series?

$$\lim_{n\to\infty} A_n = \lim_{n\to\infty} a^n \begin{cases} = 0 & \text{for } |a| < 0 \\ = 1 & \text{for } a = 1 \\ = \infty & \text{for } a > 1 \\ \text{no limit for } a \leq -1 \end{cases}$$

$$\lim_{n\to\infty} S_n = \lim_{n\to\infty} \sum_{m=0}^{n} a^m \begin{cases} = \frac{1}{1-a} & \text{for } |a| < 1 \\ \to \infty & \text{for } a \geq 1 \\ \text{no limit for } a \leq -1 \end{cases}.$$

For $a > 1$ the terms of the geometric sequence grow continuously, thus neither the sequence not the resulting series has a finite limit. For $a = 1$ the terms of the sequence are constant; the partial sums of the series correspond to the sequence of the natural numbers and thus the series is divergent.

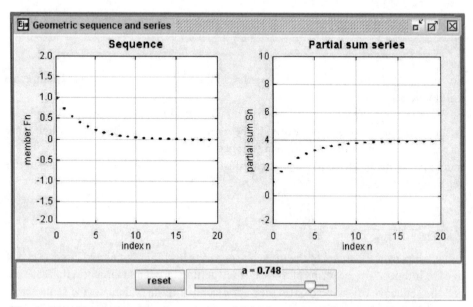

Figure 4.1a. Simulation. The first window shows the terms of the geometric sequence, the second window the partial sums of the geometric series as a function of N, with the red line as limit, provided the limit exists within the shown range of ordinates.

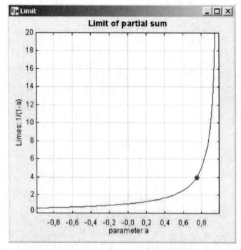

Figure 4.1b. Simulation. The third window shows the limit of the series as a function of a for $|a| < 1$. The red point marks the value of a chosen with the slide control.

For $0 < a < 1$ the terms of the sequence are getting smaller and smaller and their limit is zero. The series converges to the limit $1/(1 - a)$, which is larger than 1.

For $-1 < a < 0$ the terms of the sequence are getting smaller and smaller while changing sign and the series is convergent with the limit $1/(1 + |a|)$, which is smaller than 1.

For $a = -1$ the sequence alternates between 1 and -1 and the partial sums $(1 - 1 + 1 - 1 \pm \cdots)$ is either 1 or 0 depending on the index. Therefore no limit exists.

For $a < -1$ the terms of the sequence, as well as the partial sums, have alternating signs while growing in absolute value. Their absolute values go to infinity. Therefore the sequence and series themselves do not have a limit.

The simulation of Figure 4.1a shows the behavior of the geometric sequence and series as function of the parameter a, which can be adjusted with a slider.

What are the conditions for a series in order for it to have a limit? Obviously the terms of the associated sequence must converge to 0. That is a *necessary*, but not yet a *sufficient* condition. An example illustrating the difference is the *harmonic* series:

$$\text{harmonic series} \quad A = 1, \frac{1}{2}, \frac{1}{3}, \frac{1}{4}, \frac{1}{5}, \frac{1}{6}, \ldots$$

$$A_1 = 1; \; A_n = \frac{1}{n}; \; \lim_{n \to \infty} A_n = \lim_{n \to \infty} \frac{1}{n} = 0$$

$$S_n = 1 + \frac{1}{2} + \frac{1}{3} + \frac{1}{4} + \frac{1}{5} + \frac{1}{6} + \cdots \to \infty.$$

While the terms of the sequence converge to 0, the series grows without limit and thus does not have a limiting value.

This most easily becomes evident if one compares the harmonic series with a series that obviously diverges, and whose suitably grouped terms are smaller than or equal to those of the harmonic sequence:

$$S_{\text{harmonic}} = 1 + \frac{1}{2} + \left(\frac{1}{3} + \frac{1}{4}\right) + \left(\frac{1}{5} + \frac{1}{6} + \frac{1}{7} + \frac{1}{8}\right)$$

$$+ \left(\frac{1}{9} + \frac{1}{10} + \frac{1}{11} + \frac{1}{12} + \frac{1}{13} + \frac{1}{14} + \frac{1}{15} + \frac{1}{16}\right) + \cdots$$

$$S_{\text{comparison}} = \frac{1}{2} + \frac{1}{2} + \left(\frac{1}{4} + \frac{1}{4}\right) + \left(\frac{1}{8} + \frac{1}{8} + \frac{1}{8} + \frac{1}{8}\right)$$

$$+ \left(\frac{1}{16} + \frac{1}{16} + \frac{1}{16} + \frac{1}{16} + \frac{1}{16} + \frac{1}{16} + \frac{1}{16} + \frac{1}{16}\right) + \cdots$$

$$S_{\text{comparison}} = \frac{1}{2} + \frac{1}{2} + \frac{1}{2} + \frac{1}{2} + \frac{1}{2} + \cdots \to \infty$$

$$S_{\text{harmonic}} > S_{\text{comparison}} \Rightarrow S_{\text{harmonic}} \to \infty.$$

Thus, the terms of the harmonic sequence do not converge sufficiently strongly to zero to ensure convergence.

A sufficient criterion for convergence is that the ratio of successive terms of the sequence is smaller than 1 for $n \to \infty$ (quotient criterion of *d'Alembert*). For the two series we have:

$$\text{harmonic series} \quad \frac{A_{n+1}}{A_n} = \frac{n}{n+1}; \quad \lim_{n \to \infty} \frac{n}{n+1} = 1$$

$$\text{geometric series} \quad \frac{A_{n+1}}{A_n} = \frac{a^{n+1}}{a^n} = a; \quad \lim_{x \to \infty} a = a < 1 \text{ for } a < 1.$$

While the consecutive terms of the geometric sequence decay for $a < 1$ in a constant proportion for the geometric series, the terms of the harmonic sequence keep on decaying but, in the limit of $n \to \infty$, consecutive terms are becoming "equal".

Learning about *Archimede's* calculus of π, it must have come as a surprise to antique philosophers that an infinite number of zeros can be a well defined, finite number (sidelength $0 \cdot \infty$ number of sides of the inscribed polygon, where both 0 and ∞ are limits of an infinite series). The more difficult it is to understand that the sum of an infinite number of elements, of which *none is identical to zero*, can be a finite number. This was the base of *Xenon's* paradox, and even today one should carefully reflect about it to understand limits in depth.

4.3 Fibonacci sequence

A particularly interesting sequence of natural numbers is called after its early discoverer *Leonardo Fibonacci* (ca. 1200). It is created by defining each terms as the sum of its two predecessors. Thus the formation law reads:

Fibon

$$A_0 = 0; \; A_1 = 1$$
$$A_{n+2} = A_n + A_{n+1} \quad \text{for } n \geq 0.$$

The first 25 numbers in the sequence are:

0; 1; 1; 2; 3; 5; 8; 13; 21; 34; 55; 89; 144; 233; 377; 610; 987;
1597; 2584; 4181; 6765; 10946; 17711; 28657; 46368; 75025.

The ratio A_n/A_{n-1} of consecutive terms converges very quickly to the irrational value of the *golden mean*. (In art, the golden mean is a criterion for the balance of proportions: two dimension adhere to the golden mean if the ratio of the larger one to the smaller one is the same as the ratio of the sum of both to the larger one.)

$$A_n/A_{n-1} \to \Phi = 1.618033988\ldots.$$

The first values, which can be easily obtained with an *Excel* spreadsheet are:

$$1.0; \quad 2.0; \quad 1.5; \quad 1.6666666667; \quad 1.6; \quad 1.625; \quad 1.6153846154;$$

$$1.6190476190; \quad 1.6176470588; \quad 1.6181818182; \quad 1.6179775281;$$

$$1.6180555556; \quad 1.6180257511; \quad 1.6180371353; \quad 1.6180327869;$$

$$1.6180344478; \quad 1.6180338134; \quad 1.6180340557; \quad 1.6180339632.$$

It is evident that the differences of consecutive terms to Φ alternate in sign. In this sense, the approximation to the golden mean occurs in an oscillating manner.

This ratio can also be represented as a continued fraction with $n-1$ fractions (Please try this out for the first few terms!):

$$A_n/A_{n-1} = 1 + \cfrac{1}{1 + \cfrac{1}{1 + \cfrac{1}{1 + \cfrac{1}{1 + \cfrac{1}{1 + \frac{1}{1 + \dots}}}}}} \to \Phi.$$

From this one easily obtains that $\Phi = \dfrac{2}{\sqrt{5}-1}$ as positive root of the equation

$$\Phi = 1 + \frac{1}{\Phi} \to \Phi^2 - \Phi - 1 = 0.$$

For the exponential sequence we have, from the first term onwards:

$$A_n/A_{n-1} = \frac{e^n}{e^{n-1}} = e = 2.718\ldots$$

While the sequence of ratios is constant from the beginning for the exponential sequence, the ratios for the Fibonacci sequence only approximate a constant value for $n \to \infty$. For large n, both sequences are obviously similar. From this analogy one can deduce that the Fibonacci sequence approximates an exponential sequence for $n \to \infty$. This is an indication that the Fibonacci sequence can describe growth processes, analogous to the exponential function.

There exist numerous arithmetic relationships between the terms of the Fibonacci sequence. In addition, there are many interesting application to problems of symmetry and growth, for which we refer to the given link.

4.4 Complex sequences and series

We now consider some examples of sequences z_n and series of complex numbers with partial sums $S_n = \sum_{m=0}^{n} z_m$.

Their simulation and visualization in the complex plane provides a deeper understanding of the arithmetic operations. It shows a wealth of surprising as well as aesthetically pleasing phenomena, whose study leads to an improved understanding of the underlying mathematical questions. The examples for real series considered above are special cases of similar complex sequences.

A sequence is convergent if and only if it possesses one *accumulation point*; an accumulation point is defined such that an arbitrarily small vicinity of the accumulation point, the accumulation interval, contains in the limit nearly all terms of the sequence.

For the sequences of real numbers that have been discussed above, the accumulation point is with respect to the one dimensional domain of the F_n or S_n. For the geometric sequence or series with the parameter $|a| < 1$, the accumulation point of the sequence is zero and the accumulation point of the series is the real number $1/(1 - a)$.

The concept of an accumulation point is especially descriptive for complex numbers, since it can be visualized as enclosed by a small circle in the complex plane.

As for the visualization of the elementary complex operations, we use two windows, of which the left shows the terms of the sequence z_n and the right shows the partial sums S_n of the series. The unit circle is marked red in both. In the left window the point z_1 (second point of the sequence) corresponding to a for both the geometric and the exponential series is shown enlarged. It can be *pulled with* the mouse in such a way that a can be easily changed in this way.

In the right window, the first term of the sequence is drawn enlarged; an accumulation point, if present, is encircled by a small green circle.

Remember that complex multiplication changes not only the absolute value but also the angle, if the imaginary component is not zero. A similar thing happens when adding the terms of the sequence. In general, sequences and series therefore develop in spiral trajectories on the complex plane when their terms are generated by complex multiplication.

The description in the text can be kept short, since the simulation includes a description window with several pages, of which one contains instructions for systematic experiments.

The simulation calculates 1000 terms of the sequence. For strong convergence, many points coincide close to the accumulation point, such that only a few points can be seen separately on the screen.

4.4.1 Complex geometric sequence and series

The terms of the complex geometric sequence are created in analogy to the real case with the rule:

$$z_0 = 1$$

$$z_{n+1} = z_n \cdot a; \ n \geq 0 \rightarrow z_n = a^n.$$

Here z_n is the nth term of the sequence. The parameter a is a complex number. The terms are thus given by $1, a, a^2, a^3, a^4, \ldots$; the first term z_0 is always equal to one independent of a.

The complex geometric series is created via continuous addition of the terms of the complex sequence. Its partial sums are:

$$Sn = \sum_{m=0}^{n} a^m; \quad S_n = 1 + a^1 + a^2 + a^3 \cdots + a^n.$$

The first partial sum ($n = 0$) is again, independent of a, always 1.

In the simulation shown in Figure 4.2 you can move the point a (the second point in the sequence) in the left complex plane with the mouse and observe the effect on the terms of the sequence on the left-hand plane and on the partial sums of the complex series on the right-hand plane.

The simulation is started via the ctrl key and clicking on *Simulation*. The complex geometric series converges if the absolute value of a is smaller than 1, i.e. if a lies inside of the thin red unit circle that is drawn in the left-hand plane.

In the case of convergence the limit of the series is:

$$\lim_{n\to\infty} S_n = \lim_{n\to\infty} \sum_{m=0}^{n} a^m = \frac{1}{1-a}.$$

It is situated in the center of the green accumulation circle drawn in the right window.

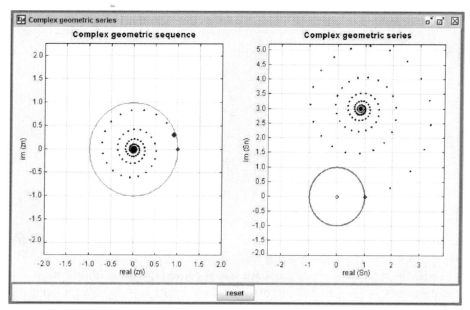

Figure 4.2. Simulation. The left window shows the terms of the geometric sequence, the second the partial sums of the series. The first point is 1 in both cases. The second point is a; it is enlarged and circled in red. Pulling this point with the mouse allows a to be changed.

For $|a| > 1$ the series diverges. The unit circle becomes smaller and smaller in the growing domain of coordinates and the series runs away along a spiral to infinity.

The case of the real geometric series is obtained as special case of the complex series if the point a is moved along the real axis. To look at the situation in more detail you can maximize the simulation window to full screen size. On the inner boundary of the unit circle, the convergence can be so slow that 1000 terms are not sufficient to nearly reach the limit. This can lead to very interesting geometrical patterns.

4.4.2 Complex exponential sequence and exponential series

The terms of the complex exponential sequence are created with the following rule:

exponential sequence: $\qquad\qquad\qquad z_{n+1} = z_n \cdot \dfrac{a}{n}$

geometric sequence for comparison: $\quad z_{n+1} = z_n \cdot a.$

Here z_n is the nth term of the sequence. The parameter a can be a complex number. We again have $z_0 = 1$

The terms thus have the form:

$$1, \frac{a}{1}, \frac{a^2}{1 \cdot 2}, \frac{a^3}{1 \cdot 2 \cdot 3}, \frac{a^4}{1 \cdot 2 \cdot 3 \cdot 4} \cdots z_n = \frac{a^n}{n!}$$

n factorial: $\quad n! = 1 \cdot 2 \cdot 3 \cdot 4 \cdots n; \quad 0! = 1; 1! = 1.$

The complex exponential series is created via continued addition of the terms of the complex exponential sequence. Thus its partial sums are:

$$S_n = \sum_{m=0}^{n} \frac{a^m}{m!}$$

$$S_n = \frac{1}{0!} + \frac{a}{1!} + \frac{a^2}{2!} + \cdots + \frac{a^n}{n!} = 1 + a + \frac{a^2}{2} + \cdots + \frac{a^n}{n!}$$

$$S_0 = 1.$$

The complex sequence and series are shown in Figure 4.3.

The case of the *real* exponential series is obtained as a special case of the complex series, if the point a is chosen on the real axis.

The terms of the exponential sequence always converge to zero. The exponential series converges for every finite value of a. The convergence is so fast that the simulation window will only show a few of the 1000 calculated terms separately.

Why does the exponential series converge so quickly as compared with the geometric series, and for any arbitrary z? In order to understand this we again consider the

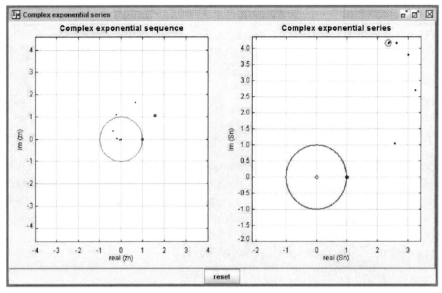

Figure 4.3. Simulation. In the simulation for the complex exponential series, which again calculates 1000 points, a (red point) can be changed with the mouse in the left window and one can see the effect on the terms of the sequence z_n, and in the right window the effect on the partial sums S_n of the complex series. The zeroth terms of both sequences are 1 and thus are situated on the red unit circle.

ratio of consecutive terms of both sequences:

$$\text{geometric sequence} \qquad \frac{z_{n+1}}{z_n} = a$$

$$\text{exponential sequence} \qquad \frac{z_{n+1}}{z_n} = \frac{a}{n}.$$

For the geometric series we must have $a < 1$ in order for the terms of the sequence to decrease, and this applies to all terms. For the exponential series, the initial terms of the series can even increase significantly! Irrespective of the size of $|a|$, as long as it is finite, there is always an index n from which the terms get smaller and smaller in absolute value, independent of the chosen a-value. Therefore we have $z_n \to 0$ irrespective of the chosen value of a.

One can easily generalize the statement concerning the convergence of the exponential series: we are given a bounded sequence B_n of numbers, which are multiplied with the respective terms of the exponential sequence. The new series is thus given by:

$$S = \sum_{m=0}^{\infty} B_m \frac{a^m}{m!} = B_0 + B_1 a + B_2 \frac{a^2}{1 \cdot 2} + \cdots + B_m \frac{a^m}{m!} + \cdots$$

$|B_m| < q$ with q real, positive number $\rightarrow$

$$|S| < q \sum_{m=0}^{\infty} \frac{a^m}{m!}; \quad S \text{ is convergent, since } \sum_{m=0}^{\infty} \frac{a^m}{m!} \text{ converges.}$$

If the absolute values of the coefficients B_m stay smaller than an arbitrary large real number q, i.e. the sequence B_m does not diverge, then the series converges, since it is smaller than the convergent exponential function multiplied by a real number. This shows how strongly the exponential series itself converges. We will later apply this result to the convergence of the Taylor expansion.

For the limit of the exponential series we have:

$$\lim_{n \to \infty} S_n = \lim_{n \to \infty} \sum_{m=0}^{n} \frac{a^m}{m!} = e^a; \quad e = 2.71828 \ldots \text{ Euler's number.}$$

If $a = 1$ one obtains

$$e = \lim_{n \to \infty} \sum_{m=0}^{n} \frac{1}{m!} = 1 + \frac{1}{2} + \frac{1}{6} + \frac{1}{24} + \cdots .$$

If one moves a in the simulation parallel to the imaginary axis, the limit of the series moves periodically on a circle around the origin. Thus one obtains "experimentally" the famous Euler formula:

With $a = x + iy$

$e^a = e^x e^{iy} = e^x (\cos y + i \sin y)$.

For $x = 0 \rightarrow e^{iy} = \cos y + i \sin y$

$$e^{iy} = 1 + iy - \frac{y^2}{2!} - i\frac{y^3}{3!} + \frac{y^4}{4!} + i\frac{y^5}{5!} \mp \cdots$$

$$= 1 - \frac{y^2}{2!} + \frac{y^4}{4!} \mp \cdots + i\left(y - \frac{y^3}{3!} + \frac{y^5}{5!} \mp \cdots \right)$$

$$\rightarrow \cos y = \sum_{n=0}^{\infty} (-1)^n \frac{y^{2n}}{(2n)!}; \quad \sin y = \sum_{n=0}^{\infty} (-1)^n \frac{y^{2n+1}}{(2n + 1)!}.$$

Euler's formula is useful for the easy derivation of relationships involving trigonometric functions. Two examples:

we are looking for: $\cos 2\varphi$, $\sin 2\varphi$

$$\cos 2\varphi + i \sin 2\varphi = e^{i2\varphi} = (e^{i\varphi})^2 \rightarrow$$

$$\cos 2\varphi + i \sin 2\varphi = (\cos \varphi + i \sin \varphi)(\cos \varphi + i \sin \varphi)$$

$$= (\cos \varphi)^2 - (\sin \varphi)^2 + i \, 2 \cos \varphi \sin \varphi$$

$$\rightarrow \cos 2\varphi = (\cos \varphi)^2 - (\sin \varphi)^2$$

$$\sin 2\varphi = 2 \cos \varphi \sin \varphi$$

we would like to evaluate: $\cos(\varphi_1 + \varphi_2)$, $\sin(\varphi_1 + \varphi_2)$

$$\cos(\varphi_1 + \varphi_2) + i \sin(\varphi_1 + \varphi_2) = e^{i(\varphi_1 + \varphi_2)} = e^{i\varphi_1} e^{i\varphi_2}$$

$$= (\cos \varphi_1 + i \sin \varphi_1)(\cos \varphi_2 + i \sin \varphi_2)$$

$$\rightarrow \cos(\varphi_1 + \varphi_2) = \cos \varphi_1 \cos \varphi_2 - \sin \varphi_1 \sin \varphi_2$$

$$\sin(\varphi_1 + \varphi_2) = \cos \varphi_1 \sin \varphi_2 + \sin \varphi_1 \cos \varphi_2.$$

Whenever one works with oscillations, i.e. with trigonometric functions, for example in optics and electronics, the use of complex numbers has many practical advantages.

From Euler's formula we obtain an elegant approximation formula for π if we put $y = \pi$ (you may convince yourself in the the simulation that the exponential function indeed yields -1 for $z = i\pi$).

$$y = \pi \rightarrow e^{i\pi} = \cos \pi + i \sin \pi = -1 + i \cdot 0 = -1$$

$$e^{i\pi} = 1 + i\pi - \frac{\pi^2}{2!} - \frac{i\pi^3}{3!} + \frac{\pi^4}{4!} + i\frac{\pi^5}{5!} - \frac{\pi^6}{6!} - i\frac{\pi^7}{7!} \cdots = -1$$

Separation in real and imaginary parts $\rightarrow$

$$\text{Re} \rightarrow 2 = \frac{\pi^2}{2!} - \frac{\pi^4}{4!} + \frac{\pi^6}{6!} - \frac{\pi^8}{8!} + \frac{\pi^{10}}{10!} \mp \cdots$$

$$\text{Im} \rightarrow 0 = \pi - \frac{\pi^3}{3!} + \frac{\pi^5}{5!} - \frac{\pi^7}{7!} + \frac{\pi^9}{9!} - \frac{\pi^{11}}{11!} \pm \cdots$$

$$\pi \neq 0 \rightarrow 0 = 1 - \frac{\pi^2}{3!} + \frac{\pi^4}{5!} - \frac{\pi^6}{7!} + \frac{\pi^8}{9!} - \frac{\pi^{10}}{11!} \pm \cdots .$$

The equations are polynomials in π^2. Neglecting all higher powers, the two series yield in zeroth order the solutions $\sqrt[2]{4} = 2$; $\sqrt[2]{6} = 2.449\ldots$. Using iterative methods of solution, for example fixed point iteration in *Excel*, one obtains the following quickly converging values, which are listed below together with the highest powers taken into account:

Approximations using the last equations (in brackets the highest power of π kept):
$(\pi^2) \rightarrow \sqrt{6} = 2.4$; $(\pi^6) \rightarrow 3.078$; $(\pi^{10}) \rightarrow 3.1411$; $(\pi^{14}) \rightarrow 3.1415920$.

Subtraction of both equations leads to a series that converges even faster, with the zeroth order solution for the order π^4: $\sqrt[4]{60} = 2.78$.

4.5 Influence of limited accuracy of measurements and nonlinearity

4.5.1 Numbers in mathematics and physics

In the domain of abstract mathematics the following relation applies exactly: $2 \cdot 2 = 4$. Exactly means that, if one were to write all numbers as decimal numbers, there would be an infinite number of zeros after the dot.

There is an old joke about the natural scientist who solved the same problem on his slide rule and obtained $2 \cdot 2 = 3.96$. Where is the difference?

In mathematics, numbers and the operations between them are defined in such a way that the repetition of the same procedure yields exactly the same result. When transferring the mathematical rules for operations to the domain of the natural sciences, there is often an unspoken assumption that not only are the operations exact and unchangeable, but also the quantities to which the operations are applied as numbers.

This is, however, not the case. When repeating an experiment in the natural sciences, one cannot assume that the natural situation in which the experiment takes place stays exactly the same;[13] above all, one has to take into account that there are limits to the accuracy of a measurement; that, even assuming fictitious equal conditions, the measured values describing the result will not be identical in a mathematical sense.

The achievable relative accuracies of measurement are often in the range of 10^{-6} to 10^{-2} with a corresponding inaccuracy of the single measurement. The highest accuracy nowadays can be reached in laser spectroscopy for the measurement of frequencies, with a relative error of 10^{-16}. For 2 consecutive measurements, one has to expect a maximum difference of this order between the results of the measurement. The result of a single measurement is only known with this accuracy.

It is the essential purpose of mathematical physical models to forecast, from the knowledge of the current state, events in the future, or to reproduce from this knowledge the past. That is the content of every formula in which the time t appears. The limited accuracy of measurements puts a natural limit on this goal.

The predictability does, however, not only depend on the accuracy for the measurements of numbers, but also on the mathematical operation that is applied to them. For a formula, such as $a = (b + b \cdot F)^n$, where b is the "true" error-free value and F

---

13 Already, the philosopher Heraclitus (around 500 BC) of antiquity realized that one "cannot bathe twice in the *same* river" *Panta rhei* (everything flows), all states are unique.

is the relative measurement error, the result depends, in addition, on the parameter n, which describes the relationship between a and b.

For an error that is small relative to the measured value, we can estimate the effect of n easily:

$$a = (b + b \cdot F)^n = b^n(1 + F)^n = b^n \sum_0^n \binom{n}{k} F^n$$

$$n = 1 \rightarrow a = b(1 + F) \quad \text{linear relationship}$$

$$1 < n \ll \left(\frac{1}{F}\right) \rightarrow a \approx b^n(1 + nF).$$

For a *linear* relationship ($n = 1$) and an accuracy of 1%, the result also has an uncertainty of 1%. In the 18th century the thinking in the philosophy of natural sciences was dominated by the conviction that the future could be forecast without limit given sufficiently accurate knowledge about the current state (Laplace's Demon);[14] this corresponds to *linear thinking*.

For *nonlinear* operations, the dependence of the results from the measurement error is also nonlinear. For the power function $a = b^n(1 + F)^n$, with $n > 1$ used in the example Figure 4.4 shows the dependence of the total error on the measurement error for increasing powers of n.

The maximum relative total error[15] grows with the power n; for the relatively small error $< 10\%$ the growth is nearly a linear function of the power; a measurement error of 1% leads, for the 10th power, to a total error of slightly over 10%.

So what? Then one has to make more accurate measurements!

However, many important and fundamental functions of physics, such as the trigonometric function, the exponential function and $1/r$-dependencies on the radius, are highly nonlinear, if one does not restrict them to a small region of values.

Even relatively small nonlinearities become important if sequences are calculated for which the next term depends on the previous term and its accuracy. This is, for example, the case if differential equations have to be solved numerically, in which hundreds of individual calculations may easily be concatenated.

Thus in view of the limited accuracy of measurements, one has to be careful for what time horizon one makes predictions with mathematical models based on measured initial data, and one has to take nonlinearities in the model used into account.

---

14 Marquis Pierre Simon de Laplace: "We may regard the present state of the universe as the effect of its past and the cause of its future. An intellect which at any given moment knew all of the forces that animate nature and the mutual positions of the beings that compose it, if this intellect were vast enough to submit the data to analysis, could condense into a single formula the movement of the greatest bodies of the universe and that of the lightest atom; for such an intellect nothing could be uncertain and the future just like the past would be present before its eyes." (*Essai philosophique sue les probabilités* 1814, Preface).

15 For simplicity we discuss the maximum error and do not discuss the statistically relevant mean error, which would not lead to any other conclusion

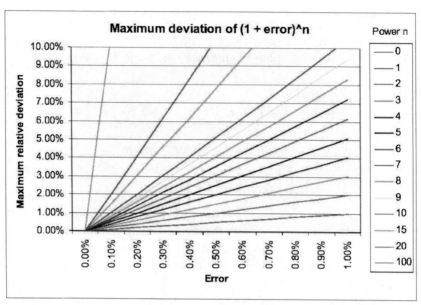

Figure 4.4. Measurement error and its effect for the relationship between measured quantity and result for a power function. The measurement error is plotted on the abscissa and the error of the result on the ordinate. The parameter is the power n.

In addition, one must not lose sight of how accurately the model used describes the reality.

When using computational models, this caution is easily lost, since the computer treats models and numbers within the limits of its computational accuracy as if they were exact in the mathematical sense. One also uses the exactly same initial values for repeated calculations.

4.5.2 Real sequence with nonlinear creation law: Logistic sequence

Even in the abstract mathematical domain, nonlinear functions produce unexpected and sometimes bizarre results. This has nothing to do with limited accuracy, but it lies in the nature of the subject. However, the resulting dependence of the calculated numbers on the initial values is so extreme that fundamental limits are imposed on transferring these models to physics or technology. An in-depth discussion of these matters can be found in *Grossmann's* essay in *Physik im 21. Jahrhundert* (detailed in the preface). We will visualize two of these phenomena using number sequences: *bifurcation* and *fractals*. The first example is concerned with a real sequence, the second one with a complex sequence.

For the sequences with a free parameter a considered so far, the creation law for the terms of the sequence depended *linearly* on a parameter:

$$\text{geometric sequence } \frac{A_n}{A_{n-1}} = a; \quad \text{exponential sequence } \frac{z_n}{z_{n-1}} = \frac{a}{n}.$$

The behavior of the sequences and the resulting series was relatively simple and clear. Is this still the case if the creation law is nonlinear? As an example we chose the so-called *logistic sequence*. This is a model for the development of a population of plants or animals under constant environmental conditions from an arbitrary initial state x_0 for a given reproduction rate. (In agreement with the notation in the literature we choose the letter x for the terms of the sequence):

$$x_{n+1} = 4r x_n (1 - x_n) = 4r(x_n - x_n^2).$$

The factor 4 scales the sequence in such a manner that, for parameter values with $0 \le r \le 1$ all terms of the sequence satisfy $0 \le x_n \le 1$.

The logistic sequence assumes firstly that the population in the next generation is proportional to the already present population. This alone would lead to unbounded, exponential growth. However, at the same time a death rate that depends quadratically on the population already present is assumed $(-4r x^2)$; note that, due to the definitions given above, we have $x_n < 1$ and therefore always $x_n^2 < x_n$.

The question that arises is: does the population for a given growth parameter under equal conditions approach a stable limit for an infinite number of generations, and how does this limit depend on the initial value x_0 and the growth parameter r?

Population growth only occurs if $x_{n+1} > x_n$, which means for $r > 1/(4(1 - x_n))$. Since $0 \le x_n \le 1$, all populations with $r < 0.25$ decay to zero independent of the initial value. For larger growth rates, i.e. for $r > 0.25$, one would therefore expect that the population would grow up to an asymptotic value larger than zero, or if initially larger, would decay to this asymptotic value.

In the simulation in Figure 4.5, r is increased consecutively by 0.001 in the interval $0 \le r \le 1$. Now a loop calculates 2000 terms of the sequence for constant r. Then one proceeds in steps of 0.001 to the next value of r until $r = 1$ is reached. Each calculation starts with a random value $0 < x_1 < 1$ for the initial value. The first terms of the sequence still depend on the initial value; therefore the first 999 iterations are not shown in the figure. The iterations 1000 to 2000 are mapped to points in the figure.

For $r < 0.75$ these points coincide so closely that a limit line as function of r is seen, comparable to $1/(1 - a)$ for the geometric sequence. Different initial values do not lead to discernable differences for the shown terms of the sequence with high indices.

For growth rates $r > 0.75$ the asymptotic orbit develops two branches (*bifurcation*), which means the iteration creates two different accumulation points. This bifurcation repeats itself until there are finally no accumulation points visible. Since 1000 iterations are shown, there could be up to 1000 values for a given r. Thus in this region

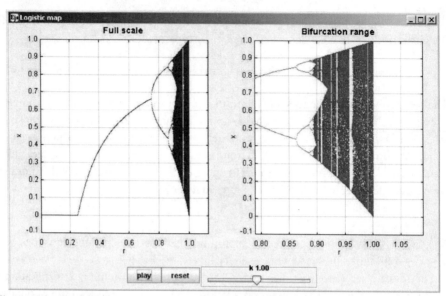

Figure 4.5. Simulation. Variation of the logistic sequence with adjustable power k (in the picture a standard sequence with $k = 1$). On the left-hand side the total range of the abscissa is shown, on the right-hand side a stretched region after the first bifurcation. The *play* button starts the animated calculation and the *reset* button resets the animation.

there cannot exist a unique limit. Surprisingly, some regions of r follow that show fewer accumulation points. The determining factor for the growth limitation is the growth rate r.

Bifurcation behavior does not depend on the growth limiting factor being exactly $1 - x_n$. Essential is the *nonlinearity* of the operation $x_n - x_n^2$. To make this experimentally accessible, a generalized factor $(1 - x_n^k)$ with $k > 0$ was chosen:

$$x_{n+1} = 4rx_n(1 - x_n^k).$$

In the simulation example you can change k *after resetting* with the slider between 0.1 and 2. the default value is 1, which leads to the usual quadratic operation.

The left window shows for the classical case ($k = 1$) the total orbit as function of r; the right window shows the bifurcation in larger resolution. For $k \neq 1$, the general character of a bifurcation stays the same, but the characteristic parameter values are moved relative to the logistic sequence and the abscissa range is adjusted accordingly.

For a more accurate viewing, the simulation window can be maximized.

In the total picture of the logistic sequence, compactified structures of accumulation points appear, which are not visible if the number of iterations shown is so large that the pixel resolution of the screen does not reveal any holes and if the resolution along the x axis is small. The simulation in Figure 4.6 therefore shows the structure of the picture with a very large horizontal resolution (~ 1000 points in the shown r-interval

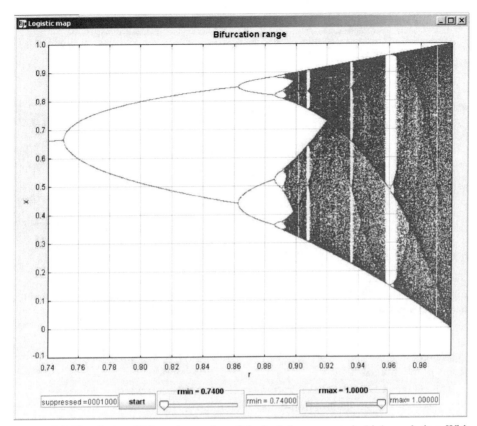

Figure 4.6. Simulation. Bifurcation region of the logistic sequence in high resolution. With the left slider for the beginning and the right slider for the end of the shown abscissa, the region can be adjusted such that particular regions can be stretched a great deal. This choice can be made very accurately by entering numbers in the fields rmin and rmax.

and a limited number of 250 iterations shown. Please maximize the window before the start of the simulation to full screen size in order to see the details. The lower and upper boundary of the r-range can be adjusted with sliders.

What is the reason for the strange behavior which becomes deterministically chaotic for large values of r? This becomes evident if one extends the simulation to show the terms of the sequence with low indices, which are suppressed in the above presentation to elucidate the limit of the sequence.

Thus one can consider individual terms of the sequence and investigate how the bifurcation results from jumping between terms with different indices.

The simulation in Figure 4.7a and Figure 4.7b, which is a real *mathematical experimentation kit*, calculates an adjustable number of terms. With the slider, the constant initial value x_0 of the sequence for a total r-scan can be adjusted. In the image an

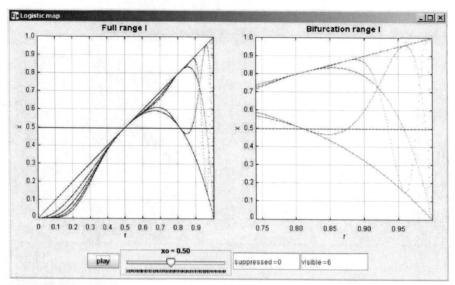

Figure 4.7a. Simulation. Individual terms of the logistic sequence, x_0: initial value; first text field: number of supressed iterations; second text field: number of iterations shown.

adjustable number of terms is shown. One can also choose the number of suppressed terms in the image.

Thus you can view the first iterations as shown in the left window of Figure 4.7a or you can look at a single iteration with a high index as in Figure 4.7b.

If one considers for example the first six terms x_0 to x_5 (*suppressed* $= 0$, *shown* $=$ 6) of the sequence as shown in Figure 4.7a, one recognizes the different terms from the increasing degree of the polynomial (The initial value as zeroth term is a straight line, the first term a line with positive slope). If you use different initial values, the images show differences in their detail. In the lower region of r one recognizes, however, how already the lower iterations approach a limiting curve. The higher iterations are then superimposed in such a way that there are nearly empty regions close to points that nearly coincide. Here the bifurcations can be found at higher indices. For the higher iterations the influence of different initial values becomes smaller and smaller.

If one shows for large indices only one term x_n, such as in Figure 4.7b, no bifurcation can be seen, but the curve shows kinks at the bifurcation points. If one increases the index by one, the kinks turn in the opposite direction. If one shows two terms x_n, x_{n+1} with consecutive index, one sees the first bifurcation. This bifurcation is thus the superposition of two r-scans with indices whose difference is 1.

Studying the conditions for lower indices one realizes that the bifurcation is caused by the change from even to odd powers that determine the individual terms.

Thus the deeper reason for the strange topology is that, for suitably defined polynomials of high order, limited regions exist, for which different orders and initial values

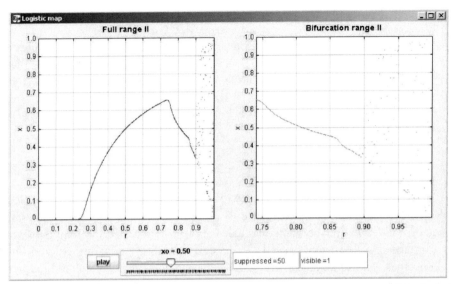

Figure 4.7b. Example from the simulation in Figure 4.7a. The 51st iteration of the logistic sequence is shown.

lead to practically identical values, while in other regions the values diverge, thus *deterministic chaos* reigns. In the essay by *Siegfried Grossmann*, this is analyzed in a general sense and in detail, and we suggest that at this point you study his contribution.

Remembering the starting point of the discussion, namely that the logistic curve is a model for the development of populations, one can draw, for example, the following conclusions: For small growth rates the population converges in an oscillating manner to a constant value at which the population and resources are in equilibrium with each other. For a higher growth rate the population exceeds the value that would be compatible with the resources. Therefore the next generation reverts to a lower value, and this jumping back and forth is repeated: the system oscillates between extremes.

The essential practical conclusion is that the result of computations for a nonlinear system can depend so sensitively on parameters and progress of the calculation (the iteration index), that a forecast is only possible for a limited number of generations. If in a nonlinear model time is the essential parameter, then this is true for any forecast over time.

It is therefore part of the art of engineering to avoid regions and dependencies in which nonlinearities lead to non-predictable or non-unique results. This is no mean feat, as most physical relationships are well determined, but nonlinear.

4.5.3 Complex sequence with nonlinear creation law: Fractals

We conclude the chapter on sequences and series with an example of a complex se- [Fracta
quence with a nonlinear creation law. Such sequences lead to the aesthetically pleasing structures called fractals, of which the *Mandelbrot set* is probably the most well
known.

Its creation law reads:

$$z_{n+1} = z_n^2 + c$$

$$z_0 = 0; \quad c : \text{complex number.}$$

For every point c of the complex plane within a limited, but sufficiently large, closed
region around the origin, the sequence is calculated and it is checked to determine
whether it diverges (in the numerical calculation it is assumed that this is the case
as soon the absolute value exceeds 4; the corresponding points are colored blue),
or converges. Those points for which the sequence converges are colored red in the
graphical representation. The points that are converging to finite values (the boundary
of the red surface) constitute the *Mandelbrot set*. All points that do not belong to it
are, depending on the speed of divergence of the sequence, shown in different colors.

The interactive Figure 4.8a provides access to a slightly modified *Mandelbrot fractal*, for which the initial value z_0 can be changed via pulling the white point with the
mouse; $z_0 = 0$ gives the well known Mandelbrot set, $-2 < z < 2$ covers the region
in which convergence happens at all. Resetting leads to the initial state.

The region of the calculation can be restricted by specifying a region with the
mouse; multiple restriction makes it possible to delve into deep regions of the fractal
ramifications (see as an example Figure 4.9b).

Figure 4.8b shows the modified Mandelbrot set for $z_0 = i$.

The topologically novel situation of the fractal structure is that the boundary of a finite area is infinitely branched and shows self-similarity when delving deeper and
deeper, i.e. on all scales similar structures are visible. You will realize this when
selecting ever smaller sections.

It is not trivial to understand which mathematical relationship leads to the special
form and symmetry in the figure.

To simplify this task we generalize further, to use instead of the quadratic creation
rule an arbitrary power:

$$z_{n+1} = z_n^k + c$$

$$z_0 = 0; \quad c : \text{complex number}$$

$$k \geq 1.$$

For $k = 2$ we find for the set of c-values for which z_n does not diverge the Mandelbrot
set as discussed above.

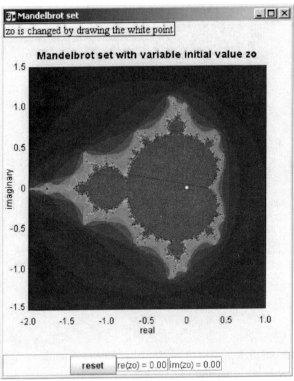

Figure 4.8a. Simulation. Modified Mandelbrot set with adjustable initial value z_0 (white point) for the iteration (in the picture we have the standard set: $z_0 = 0$). The coordinates of z_0 are shown in two output fields and can be adjusted via pulling the white point with the mouse or via entering values.

In the simulation in Figure 4.9a the power k can be changed by a slider to a rational number between 1 and 10. In the number field values that are unlimited can be entered (after input you need to press the *enter* key and must wait until the input field changes color again!). For this simulation many trigonometric functions have to be calculated, which requires a lot of effort. Thus you need to be patient after the first call or after entering a new value. Depending on the resources of your computer this calculation can take several seconds or even minutes.

Figure 4.9a shows the modified Mandelbrot set of the c-values for which the complex point sequence z_n converges for $k = 1000$. The region of convergence to nonzero nearly corresponds to the unit circle (as one expects for the geometric series), but exhibits further fractal branching at the boundary, as shown in Figure 4.9b in higher resolution.

An aesthetically particularly interesting variant of a given complex fractal is its *Julia set*. This is obtained by keeping the point c fixed in the complex plane and asking which points z in the plane lead to a divergent or convergent sequence. Thus

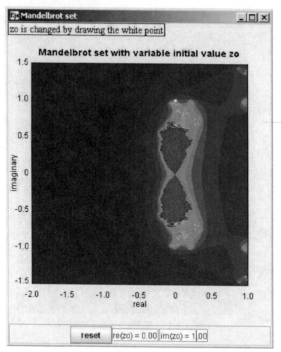

Figure 4.8b. Simulation. Modified Mandelbrot set from simulation in Figure 4.8a for $z_0 = i$.

for the Mandelbrot set and its Julia set we have:

creation law of sequence $z_{n+1} = z_n^2 + c$

Mandelbrot set:

$z_0 = $ constant $ = 0$.   For which points c does the sequence converge or diverge?

corresponding Julia set:

$c = $ constant;   For which points z does the sequence converge or diverge?

Thus one can map every point c of the Mandelbrot set to its Julia set. In the simulation of Figure 4.10 a small white point in the left window showing the Mandelbrot set can be moved with the mouse. The program calculates the corresponding Julia set, which is shown in the right-hand window. Its appearance and symmetry change in a characteristic manner if one moves c around the Mandelbrot Set. With the slider one can adjust the color shading for the diverging values.

$c = 0$ leads to the sequence $z, z^2, z^4, z^8, \ldots$, which is the geometric sequence that converges inside the unit circle and diverges outside it. The Julia set is now identical with the unit circle.

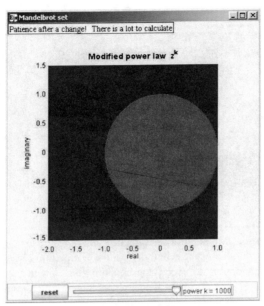

Figure 4.9a. Simulation. Modified Mandelbrot set with adjustable power (1000 in the picture). The red surface fills the unit circle nearly completely; only at the boundary some branching can be seen. You can also choose rational numbers for the power. To get the original Mandelbrot set enter the integer 2 in the number field.

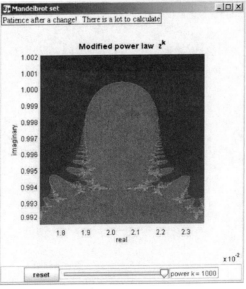

Figure 4.9b. Fractal branching of the boundary of Figure 4.9a, shown at corresponding high resolution, obtained by specifying a rectangular region with the mouse and zooming twice.

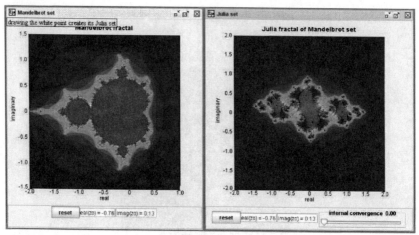

Figure 4.10. Simulation. Mandelbrot set and corresponding Julia set; the parameter c of the Julia set is adjusted by moving the white dot (in the left picture close to the upper incision between "head" and "body"). By specifying bounding boxes with the mouse one can again restrict the computation region in both fields. Using the slider the color mapping of convergence for the Julia set can be adjusted leading to a variety of color patters symbolizing the convergence.

5 Functions and their infinitesimal properties

5.1 Definition of functions

Traditionally we speak of a function $f(x)$ if every x satisfying $x_1 < x < x_2$ is <u>Function</u> mapped to another number y in a unique way; here $y = f(x)$ is the mapping prescription. An example is $y = \sin(x)$, with real numbers x and y, or $u = z^n$ and with complex numbers z, u and a real number n. For brevity one can also write $y(x)$ instead of $y = f(x)$.

In a more general manner, one can define the concept of a function by mapping each element a of a set A uniquely to an element of the set B: *the set A is mapped to the set B via the function f*:

$$B = f(A)$$

In this example, $a \in A$ can be referred to as the preimage (or inverse image) point and $b \in B$ as the image or image point.

Functions and mapping are synonymous concepts, the concept of which includes the uniqueness of the mapping.

The converse assumption is, however, not necessarily true: an image point b can have numerous *preimages $a, a', \ldots$*. For the sine function there is, for every x, a unique value $y = \sin(x)$. Due to periodicity of the sine function modulo 2π, every y can be mapped to arbitrarily many x points.

The sequences and series discussed in Chapter 4 are examples of the mapping of *discrete* numbers – that is, of functions whose domain of definition for x consists of discrete values n.

In general, the domain of definition of the variable a of a function will be *continuous* set, i.e. the set of real or complex numbers, or a limited region of one of these sets.

A function is *continuous* in the domain of definition of its preimage if the set of variables $a \in A$ is dense and, in addition, an arbitrarily small neighborhood of a_0 is mapped to a dense neighborhood of the image point b_0. Visually, this means that there are no gaps or jumps in the curve corresponding to $b(a)$.

The limit of the complex geometric series can be considered as the mapping of the continuous complex domain a inside of the unit circle to the complex plane outside of the unit circle:

$$|a| < 1$$

$$z = f(a) = \lim_{n \to \infty} \sum_{m=0}^{n} a^m = \frac{1}{1-a} \to 1 \leq z < \infty.$$

The function is continuous in its domain of definition $|a| < 1$.

5.2   Difference quotient and differential quotient

For a continuous function, the variable x can have any value within its domain of definition X. As in sequences one can define a *difference quotient* as the difference of two function values y_2 and y_1 with different values x_2 and x_1 of the independent variable.

While the difference quotient for sequences was given by the difference of two terms with an index difference of 1, i.e. $A_{n+1} - A_n$, the difference $y_1 - y_1 = f(x_2) - f(x_1)$ can be defined for an arbitrarily small difference $\Delta x = x_2 - x_1$ in the case of continuous functions.

In addition one can define a *differential quotient* as the limit for an infinitesimal distance between the variables Δx. Thus it becomes a *local* property of the function in every point x, in which such a unique value exists, i.e. at which the function is *differentiable*.

difference quotient:
$$\frac{\Delta f}{\Delta x} = \frac{f(x_2) - f(x_1)}{x_2 - x_1} = \frac{f(x_1 + \Delta x) - f(x_1)}{\Delta x};$$

differential quotient:
$$\left(\frac{df}{dx}\right)_{x_1} = \lim_{\Delta x \to 0} \left(\frac{\Delta f}{\Delta x}\right)_{x_1} = \lim_{\Delta x \to 0} \frac{f(x_1 + \Delta x) - f(x_1)}{\Delta x}.$$

For $\Delta x > 0$ we refer to a right-hand difference – or differential quotient, for $\Delta x < 0$ to a left-hand one. If both differential quotients exist and are equal, then the function is *uniquely* differentiable at this point.

If the function is uniquely differentiable in every point of its domain of definition – it is then also continuous – its differential quotient is a continuous function of the variable x, the *first derivative* of the function:

$$y'(x) = f'(x) = \frac{df}{dx}(x) = \lim_{\Delta x \to 0} \frac{f(x + \Delta x) - f(x)}{\Delta x}.$$

As shorthand one writes the first derivative as y' (*y prime*) or $f'(x)$.

If the differential quotient exists, it is a proper ratio of two numbers as their respective limits. Thus one can treat both denominator and numerator as such:

$$df = f'(x)dx$$

If the first derivative is uniquely differentiable in every point of the domain of definition one can define the second derivative, an so on:

$$y''(x) = f''(x) = \frac{df'}{dx}(x), \quad \dots, \quad y^{(n)}(x) = f^{(n)}(x) = \frac{dy^{(n-1)}}{dx}(x).$$

Of particular practical importance are continuous functions that can be arbitrarily often differentiated, also called "smooth" functions, such as the trigonometric functions.

In physics the independent variable is often the time t. For the derivative with respect to time the notation $\dot{y}$ (y dot) has been adopted in written and printed work. This is somewhat unfortunate for our purposes, since this sign cannot be directly entered on the keyboard of a PC, and also one cannot enter it as a character with two meanings (y and derivative with respect to t). In this text we stick to the notation y', even if the independent variable is the time t.

5.3 Derivatives of a few fundamental functions

5.3.1 Powers and polynomials

Normally one finds the derivatives of the most important functions in tables or one has learned them by heart in school. They are, however, very easy to calculate if one takes into account that the limit $\Delta x \to 0$ takes place and that therefore all higher powers of Δx can be neglected.

We show this in detail for the example of the second power:

$$y(x) = x^2$$

$$y(x + \Delta x) = (x + \Delta x)^2 = x^2 + 2x\Delta x + \Delta x^2$$

$$y(x + \Delta x) - y(x) = 2x\Delta x + \Delta x^2$$

$$\frac{y(x + \Delta x) - y(x)}{\Delta x} = 2x + \Delta x$$

$$y' = \lim_{\Delta x \to 0} \frac{y(x + \Delta x) - y(x)}{\Delta x} = \lim_{\Delta x \to 0} (2x + \Delta x) = 2x.$$

This can now be easily extended to arbitrary powers, if one takes into account that the second term of the polynomial $(x + \Delta x)^n = x^n + nx^{n-1}\Delta x + ax^{n-2}\Delta x^2 + bx^{n-3}\Delta x^3 + \cdots$ is $nx^{n-1}\Delta x$.

The coefficients of the further terms $a, b, c, \ldots$ do not have to be given explicitly, since all these terms vanish in the limit $\Delta x \to 0$, as they contain at least the factor Δx^2:

$$y(x) = x^n$$

$$y(x + \Delta x) = (x + \Delta x)^n = x^n + nx^{n-1}\Delta x + ax^{n-2}\Delta x^2 + cx^{n-3}\Delta x^3 + \cdots$$

$$y(x + \Delta x) - y(x) = nx^{n-1}\Delta x + ax^{n-2}\Delta x^2 + bx^{n-3}\Delta x^3 + \cdots$$

$$\frac{y(x + \Delta x) - y(x)}{\Delta x} = nx^{n-1} + \Delta x(ax^{n-2} + \cdots)$$

$$y' = \lim_{\Delta x \to 0} \frac{y(x + \Delta x) - y(x)}{\Delta x} = \lim_{\Delta x \to 0} (nx^{n-1} + \Delta x(\ldots)) = nx^{n-1}.$$

This also yields the rule for higher derivatives of powers:

$$y(x) = x^n, \quad y' = nx^{n-1}, \quad y'' = n(n-1)x^{n-2}, \quad \ldots$$

$$y^{(n)} = n(n-1)(n-2)\ldots\ldots(1) = \text{const}$$

$$y^{(n+1)} = 0.$$

The derivative of a constant c, which has by definition the same value of all values of the independent variable, is zero.

The rules obtained above also apply if the exponents are negative or rational:

$$y = x^{-n} = \frac{1}{x^n} \to y' = -nx^{-n-1} = -nx^{-(n+1)} = -\frac{n}{x^{n+1}}$$

$$y = \sqrt[3]{x} = x^{1/3} \to y' = \frac{1}{3}x^{1/3-1} = \frac{1}{3}x^{-2/3} = \frac{1}{3\sqrt[3]{x^2}}.$$

With this result it is also easy to see how the derivatives of polynomials look; for example:

$$y = 3x^5 + 4x^4 + 3x - 1$$

$$y' = 15x^4 + 16x^3 + 3$$

$$y'' = 60x^3 + 48x^2; \quad y''' = 180x^2 + 96x;$$

$$y^{(4)} = 360x + 96; \quad y^{(5)} = 360; \quad y^{(6)} = 0.$$

We have shown the formal differentiation of powers in so much detail because this also allows us to treat functions for which a series expansion containing power terms is known.

5.3.2 Exponential function

In analogy to the exponential series, we can define the exponential function for a continuous domain of the variable x. Since its series expansion consists of powers, we can obtain its derivative immediately by differentiating its individual terms according to the rule derived above.

$$e = 2.71828\ldots$$

$$y = e^x = \lim_{n \to \infty}\left(S_n = \sum_{m=0}^{n} \frac{x^m}{m!}\right) = 1 + x + \frac{x^2}{1 \cdot 2} + \frac{x^3}{1 \cdot 2 \cdot 3} + \frac{x^4}{1 \cdot 2 \cdot 3 \cdot 4} + \cdots$$

$$y' = 0 + 1 + \frac{2x}{1 \cdot 2} + \frac{3x^2}{1 \cdot 2 \cdot 3} + \frac{4x^3}{1 \cdot 2 \cdot 3 \cdot 4} + \cdots = 1 + x + \frac{x^2}{1 \cdot 2} + \frac{x^3}{1 \cdot 2 \cdot 3} + \cdots$$

$$y' = y$$

$$y'' = y' = y.$$

Thus the exponential function has the property that its derivatives and the function are identical. The above derivation also shows that the coefficient of the nth term of the exponential sequence $\frac{1}{n!}$ is given by the reciprocal of the nth derivative of its respective power:

$$y = \frac{x^n}{n!}; \quad y' = \frac{n \cdot x^{n-1}}{n!} = \frac{x^{n-1}}{(n-1)!}; \quad y^{(n)} = \frac{x^0}{0!} = 1; \quad y^{(n+1)} = 0.$$

Upon differentiation, every term assumes the form of the previous term and the constant term vanishes. This property results in the exponential function and its derivative becoming identical.

5.3.3 Trigonometric functions

In an analogous manner we can obtain the derivatives of the trigonometric functions from their series expansions. We start with the representations that we previously obtained from the complex exponential function:

$$y = \sin x = x - \frac{x^3}{3!} + \frac{x^5}{5!} \mp \cdots = \sum_{0}^{\infty} (-1)^n \frac{x^{2n+1}}{(2n+1)!}$$

$$\to y' = 1 - \frac{3x^2}{3!} + \frac{5x^4}{5!} \mp \cdots = 1 - \frac{x^2}{2!} + \frac{x^4}{4!} \mp \cdots = \sum_{0}^{\infty} (-1)^n \frac{x^{2n}}{(2n)!} = \cos x$$

$$y = \cos x = 1 - \frac{x^2}{2!} + \frac{x^4}{4!} \mp \cdots = \sum_{0}^{\infty} (-1)^n \frac{x^{2n}}{(2n)!}$$

$$\to y' = -\frac{2x}{2!} + \frac{4x^3}{4!} \mp \cdots = -\left(x - \frac{x^3}{3!} + \frac{x^5}{5!} \mp \cdots\right)$$

$$= -\sum_{0}^{\infty} (-1)^n \frac{x^{2n+1}}{(2n+1)!} = -\sin x.$$

By taking into account the signs all further derivatives can be established:

$$y = \sin x \rightarrow y' = \cos x; \qquad y'' = -\sin x; \qquad y''' = -\cos x; \qquad y'''' = \sin x$$

$$y = \cos x \rightarrow y' = -\sin x; \qquad y'' = -\cos x; \qquad y''' = \sin x; \qquad y'''' = \cos x.$$

Using these results all functions, which can be described as series expansions in terms of trigonometric functions, can be easily differentiated. These are, in the main, functions that describe periodic phenomena.

5.3.4 Rules for the differentiation of combined functions

Combined functions are easy to differentiate if one knows the derivatives of the functions that are combined. The following, immediately plausible, rules apply.

multiplicative constant c

$$y = c \cdot f(x) \rightarrow y' = c \cdot f'(x)$$

additive composition

$$y = f(x) + g(x) \rightarrow y' = f'(x) + g'(x)$$

product rule

$$y = f(x) \cdot g(x) \rightarrow y' = f'(x) \cdot g(x) + f(x) \cdot g'(x)$$

quotient rule

$$y = \frac{f(x)}{g(x)} \rightarrow y' = \frac{f'(x) \cdot g(x) - f(x) \cdot g'(x)}{(g(x))^2}$$

chain rule

$$y = f(g(x)) \rightarrow y' = f'(g(x)) \cdot g'(x)$$

example $y = \sin(x^3 + x) \rightarrow y' = \cos(x^3 + x) \cdot (3x^2 + 1).$

5.3.5 Derivatives of further fundamental functions

To be able to differentiate all "prevalent" functions formally, one needs a collection of derivatives of additional fundamental functions. We list these here without comment in the form of a table together with those obtained above. The derivatives of the hyperbolic functions at the end of the table are simply obtained from their definitions in terms of exponential functions.

$$y = x^n \rightarrow y' = nx^{n-1}$$

$$y = e^x \rightarrow y' = e^x; \; y = e^{ax} \rightarrow y' = ae^{ax}; \; y = a^x \rightarrow y' = a^x \ln a$$

$$y = \sin x \rightarrow y' = \cos x; \qquad\qquad y = \cos x \rightarrow y' = -\sin x$$

$$y = \tan x \rightarrow y' = \frac{1}{\cos^2 x}; \qquad\qquad y = \cot x \rightarrow y' = -\frac{1}{\sin^2 x}$$

$$y = \arcsin x \rightarrow y' = \frac{1}{\sqrt{1 - \sin^2 x}}; \qquad y = \arccos x \rightarrow y' = -\frac{1}{\sqrt{1 - \sin^2 x}}$$

$$y = \arctan x \rightarrow y' = \frac{1}{1 + x^2}; \qquad\qquad y = \text{arccot}\, x \rightarrow y' = -\frac{1}{1 + x^2}$$

$$y = \ln x \rightarrow y' = \frac{1}{x}; \qquad\qquad y = {}^a\log x \rightarrow y' = \frac{1}{x \ln a}$$

$$y = \sinh(x) = \frac{e^x - e^{-x}}{2} \quad\rightarrow\quad y' = \frac{e^x + e^{-x}}{2} = \cosh(x)$$

$$y = \cosh(x) = \frac{e^x + e^{-x}}{2} \quad\rightarrow\quad y' = \frac{e^x - e^{-x}}{2} = \sinh(x).$$

In contrast to the trigonometric functions $\sin(x)$ and $\cos(x)$, the derivatives of the hyperbolic functions $\sinh(x)$ and $\cosh(x)$ show no additional sign change on differentiation.

For the inverse functions of the trigonometric functions we make use of notations such as $\arccos(x)$ that are employed in mathematical texts; in Java code we use instead $acos(x)$.

5.4 Series expansion: the Taylor series

5.4.1 Coefficients of the Taylor series

In many cases it is useful to analyze instead of a function $f(x)$ a series that approximates it. This is true particularly if the series converges to the function without restrictions. Then the partial sums of the series can be considered as approximations with increasing accuracy.

For the terms of the sequence that make up the series one will use such functions in preference that can be differentiated and integrated easily. Especially suitable are series whose terms are powers or trigonometric functions of the variables. The first case leads to the *Taylor series*, whose coefficients are obtained via differentiation, which we will study more closely in the following. The second case leads to the *Fourier series*, which we will visualize after treating the integral, since its coefficients are determined via integration.

Another argument for the choice of a particular series expansion can be to use functions for the terms of the series that are particularly adapted to the symmetry of the problem that is described by the function, e.g. Bessel functions for cylindrical symmetry and spherical harmonics for point symmetry.

The Taylor series is an infinite series whose partial sums are an approximation for the function $y = f(x)$, that is exact at the point x_0 and approximate in the vicinity of $x = x_0$, and the interval for an acceptable approximation becomes larger with increasing index of the partial sum. The members of the sequence that constitutes the series are powers of the distance from the computation point $(x - x_0)$. Thus the function is approximated via a power series and the problem consists of finding the coefficients of the individual terms.

To achieve this, we first equate the function formally to a power series with terms $a_n(x - x_0)^n$ and parameters a_n. We then differentiate both sides repeatedly. After each step we put $x = x_0$. Thus all powers containing $x - x_0$ drop out from the power series for the respective derivatives and the coefficient of the remaining term can be easily obtained:

$$\text{ansatz: } f(x) = \sum_0^\infty a_n(x - x_0)^n$$

$$= a_0 + a_1(x - x_0) + a_2(x - x_0)^2 + a_3(x - x_0)^3 + \cdots$$

$$(x - x_0) = 0 \to a_0 = f(x_0)$$

$$f'(x) = a_1 + 2a_2(x - x_0) + 3a_3(x - x_0)^2 + 4a_4(x - x_0)^3 + \cdots$$

$$(x - x_0) = 0 \to a_1 = \frac{f'(x_0)}{1}$$

$$f''(x) = 1 \cdot 2a_2 + 2 \cdot 3a_3(x - x_0) + 3 \cdot 4a_4(x - x_0)^2 + \cdots$$

$$(x - x_0) = 0 \to a_2 = \frac{f''(x_0)}{1 \cdot 2}$$

$$f'''(x) = 2 \cdot 3a_3 + 3 \cdot 4 \cdot 2a_4(x - x_0)^1 + \cdots$$

$$(x - x_0) = 0 \to a_3 = \frac{f'''(x_0)}{1 \cdot 2 \cdot 3}$$

$$f^{(n)} = n!a_n + \cdots + \frac{n!}{2}(x - x_0) + \cdots \to a_n = \frac{f^{(n)}}{n!}.$$

Thus the coefficient of the nth power is proportional to the nth derivative of the function at the computation point and the factorial as a factor simply follows from differentiating the nth power. The Taylor series of the function is then with $0! = 1$, $1! = 1$ and $f^{(0)}(x_0) = f(x_0)$:

$$f(x) = \frac{f(x_0)}{0!} + \frac{f'(x_0)}{1!}(x - x_0) + \frac{f''(x_0)}{2!}(x - x_0)^2 + \frac{f'''(x_0)}{3!}(x - x_0)^3 + \cdots$$

$$\text{Taylor series: } f(x) = \sum_{n=0}^\infty f^{(n)}(x_0)\frac{(x - x_0)^n}{n!}$$

zeroth approximation: $f(x) = f(x_0)$

first approximation, in x linear: $f(x) = f(x_0) + f'(x_0)(x - x_0)$

second approximation, in x quadratic:

$$f(x) = f(x_0) + f'(x_0)(x - x_0) + \frac{f''(x_0)}{2}(x - x_0)^2.$$

While $f'(x) = df/dx(x)$ describes the slope of a differentiable function at each point x in its domain of definition, $f''(x)$ i.e. $df'/dx(x)$ describes the slope of the slope or the change of the slope of $f(x)$. The slope changes if and only if the curve $f(x)$ has a curvature. Therefore $f''(x)$ is a measure of the curvature of $f(x)$. If one identifies x with the time t and $y = f(t)$ with the distance traveled by an object during the time t, the first derivative is denoted by velocity and the second derivative is called acceleration of the object that is, at time t, at position x.

The first approximation of the Taylor series takes into account the slope of the function at the computation point, the second one in addition to its curvature. The higher approximations use higher derivatives and it makes sense to also visualize their meaning.

In the simulation of Figure 5.1 the derivatives up to the ninth order are calculated for a function that can be chosen from 9 given options and are shown as colored curves in an abscissa region that depends on the function and also may have a shifted origin. With the choice boxes at the top, the derivatives to be plotted in addition to the function can be selected; all nine are shown in the figure. For the red point, which can be moved with the mouse, the values of the local values of the derivatives are calculated anew and displayed in the number fields on the left.

The derivatives are approximated numerically as differential quotients using both neighboring points:

$$y'(x) \approx \frac{y(x + \Delta x) - y(x - \Delta x)}{2\Delta x}$$

$$y''(x) \approx \frac{y'(x + \Delta x) - y'(x - \Delta x)}{2\Delta x} \approx \frac{y(x + 2\Delta x) - 2y(x) - y(x - 2\Delta x)}{4(\Delta x)^2}$$

$$\vdots$$

You will find further details about this in the description pages of the simulation.

In many simulations contained in this book it is possible to enter formulas for func- `Parser` tions directly in mathematical notation. For the program, the functions are initially strings without meaning, which have to be interpreted by an additional program, a *parser*, and translated to Java code. This is a relatively complex process. If the function is only translated once for a simulation the required time is not of concern and one has, when using the parser for EJS, the advantage of being able to change the function or enter a new one without having to open and edit the program itself.

The determination of higher derivatives with sufficient accuracy requires a considerable computational effort. In the example of Figure 5.1 the function has to be

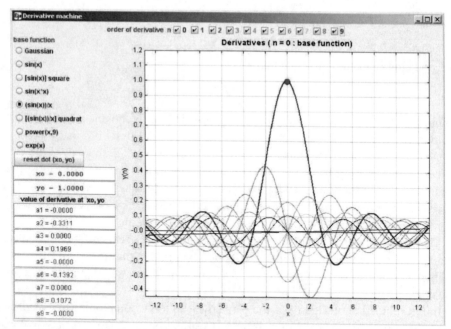

Figure 5.1. Simulation. Derivatives of a given fundamental function (blue, chosen on the left-hand side) up to the ninth order, drawn in the colors of the choice boxes above. For the red computation point, which can be moved with the mouse, the values of derivatives are given in the number fields on the left. The picture shows the derivatives of $\sin(x)/x$.

evaluated 10 000 times for one computation, which puts a strain on the computing speed of a simple PC. Therefore, the functions are predetermined in our example. If you want to analyze other functions you may open the simulation using the EJS console and change the simple Java code of the preset functions.

In the upcoming simulations of Figure 5.2 and Figure 5.3, the approximations for the derivatives are calculated once without using, and once using, the parser and you will recognize the difference in the computation speed from these examples.

Convergence of the Taylor series

It should not be taken for granted that the power series also approaches the function for values of x outside the computation point x_0. During the discussion of the exponential function, which has, as a power series, a large similarity with the Taylor series, we had, however, already established that the series also has to converge within the vicinity of the computation point if the factors attached to its term do not diverge. For the Taylor series, these factors are the derivatives in the computation point. The Taylor series converges in the neighbourhood of the computation point if in the limit $n \to \infty$ derivatives grow smaller than n.

For many functions that are important in physics, as for example *polynomials, exponential function, sine and cosine*, the domain of convergence is unlimited. With increasing order or number of terms the corresponding Taylor series approximates the original function over a larger and larger interval and the domain of small deviations becomes larger and larger. In practice one normally uses a partial sum of finite order; then the partial sum is identical to the function at the point of computation and increasingly deviates from the function with growing distance from it.

It is amusing to calculate the Taylor series of the exponential function. Since all it derivatives are equal, the Taylor series coincides with the exponential series.

The power function of degree n has non-vanishing derivatives only up to order $n + 1$. In this case, the Taylor series terminates after the $(n + 1)$th term. Its Taylor expansion is thus identical to the original function.

The trigonometric functions, however, have an unlimited number of derivatives that are repeated periodically, for example $\sin x, \cos x, - \sin x, - \cos x, \sin x, \ldots$. The approximation will become better the more terms of the series expansion are retained.

Among the possible approximation functions, the Taylor series is characterized by the fact that the coefficients can be determined from data at a computation point alone, namely all the derivatives of the function as this point. This series has the great practical advantage that its terms are powers and can therefore be easily added to and multiplied with each other, and also easily integrated and differentiated; the derivatives of the function at the computation point that appear in the coefficients are constants for the operations listed above. Therefore, in physical analysis, complex functions are often approximated by a Taylor series with a limited number of terms: *linear approximation* with two terms and *quadratic approximation* with three terms.

5.4.2 Approximation formulas for simple functions

The linear term of the Taylor series already yields approximations that are often used in practice: the derivation is shown for three basic functions; for other cases you may easily derive this yourself. You may, for example, use $x = x_0$ for the computation point and determine the next highest derivative.

Expansion around the computation point $x = 0$, applicable for $|x| \ll 1$

1.) $\quad y = \sqrt{1 + x} = (1 + x)^{\frac{1}{2}}$

$\quad y' = \dfrac{1}{2}(1 + x)^{-\frac{1}{2}} \qquad \rightarrow y \approx \sqrt{1 + 0} + \dfrac{1}{1!} \cdot \dfrac{1}{2}(1 + 0)^{-\frac{1}{2}}x = 1 + \dfrac{1}{2}x$

2.) $\quad y = \dfrac{1}{1 - x} = (1 - x)^{-1}$

$\quad y' = (1 - x)^{-2} \qquad \rightarrow y \approx 1 + x$

3.) $\quad y = \sin x; \; y' = \cos x; \qquad \rightarrow y \approx 0 + 1 \cdot x = x$

4.) $\quad y = \cos x; \; y' = - \sin x : \quad \rightarrow y \approx 1 - 0 \cdot x = 1.$

5.4.3 Derivation of formulas and errors bounds for numerical differentiation

Using the Taylor series, one can quickly obtain formulas for the numerical calculation of the first derivative y'. This also yields a measure for the respective accuracy. We show this for the linear approximation; the procedure can be easily extended to higher approximations.

We assume, in the following, that both $y(x)$ and $y(x + \Delta x)$ are known.

$$\text{Taylor series } y(x + \Delta x) = y(x) + \frac{y'(x)\Delta x}{1} + \frac{y''(x)\Delta x^2}{2}$$

$$+ \frac{y'''(x)\Delta x^3}{6} + \frac{y^{(4)}(x)\Delta x^4}{24} + \cdots \rightarrow$$

$$y'(x) = \frac{y(x + \Delta x) - y(x)}{\Delta x} - \left[\frac{y''(x)\Delta x}{2} + \frac{y'''(x)\Delta x^2}{6} + \cdots \right]$$

$$y'(x) = \frac{y(x + \Delta x) - y(x)}{\Delta x} - O(\Delta x);$$

$$\text{with } O(\Delta x) = \frac{y''(x)\Delta x}{2} + \frac{y'''(x)\Delta x^2}{6} + \cdots \approx \frac{y''(x)\Delta x}{2}.$$

The last but one line shows the usual definition for the difference quotient supplemented by the term $O(\Delta x)$ (letter O), which gives the deviation from the differential quotient due to neglecting the higher terms of the Taylor series. The deviation vanishes in the limit of $\Delta x \rightarrow 0$, since all terms contained in O depend at least linearly on Δx. For sufficiently small intervals, the higher powers of Δx can be neglected against the linear term and one obtains the important conclusion, that the procedure of differentiation according to the above formula becomes accurate *linearly* with Δx. If one halves the width of the interval, the accuracy is doubled.

Using the Taylor series one can easily derive a method with better convergence for the calculation of the derivative. We write down the Taylor series once for a point that is Δx to the right of the computation point x and once for a point that is Δx to the left of the computation point. Subtracting the two series from each other, the terms with even powers drop out:

$$[1]\; y(x + |\Delta x|) = y(x) + y'(x)|\Delta x| + \frac{y''(x)|\Delta x|^2}{2}$$

$$+ \frac{y'''(x)|\Delta x|^3}{6} + \frac{y''''(x)|\Delta x|^4}{24} + \cdots$$

$$[2]\; y(x - |\Delta x|) = y(x) - y'(x)|\Delta x| + \frac{y''(x)|\Delta x|^2}{2} - \frac{y'''(x)|\Delta x|^3}{6} \pm \cdots$$

$$[1] - [2] \rightarrow y(x + |\Delta x|) - y(x - |\Delta x|) = 2y'(x)|\Delta x| + 2\frac{y'''(x)|\Delta x|^3}{6} + \cdots$$

$$y' = \frac{y(x + |\Delta x|) - y(x - |\Delta x|)}{2|\Delta x|} - O\left(\frac{y'''(x)|\Delta x|^2}{6} + \cdots\right).$$

The formula obtained in this way converges quadratically with the width of the interval; halving the interval $|\Delta x|$ improves the accuracy by a factor of four.

One can interpret the formulas in geometric terms: the first one approximates the value of the derivative at the beginning of the interval by the slope between beginning and end. The second one approximates it by the slope between the beginning of an interval to the left and the end of an interval to the right of the computation point.

One can continue with the above procedure and thus obtain even faster converging approximation formulas; however, one then needs values of the function at more points to calculate the differential quotient. Therefore, one often sticks to the above approximation with quadratic convergence.

5.4.4 Interactive visualization of Taylor expansions

In the following we consider two simulations for visualizing Taylor expansions. The first, Figure 5.2, uses the same setup that was employed for the calculation of deriva-

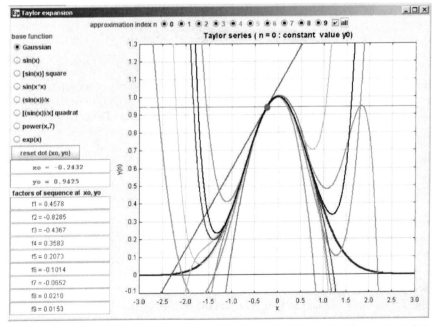

Figure 5.2. Simulation. Taylor expansions of the Gaussian (blue, selection on the left-hand side) from zeroth to ninth order around the adjustable red computation point. The Taylor coefficients f_n can be read on the left.

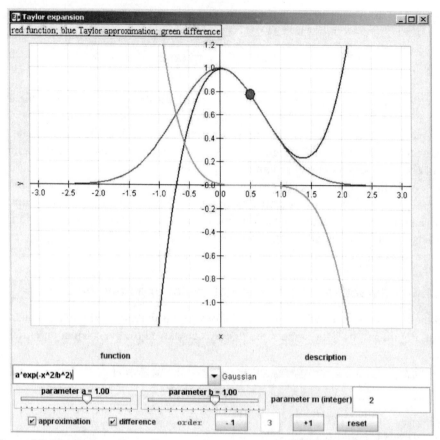

Figure 5.3. Simulation. Approximation of a function in the vicinity of an adjustable computation point via partial sums of the Taylor series; in the figure, the Gaussian is drawn in red,
the third degree approximation in blue and the deviation in green. The computation point in
magenta can be pulled with the mouse and the degree of approximation can be increased or
decreased by one with the +1 and −1 keys. Two free parameters a and b can be continuously
adjusted with sliders and a third integer parameter m can be changed in the number field. The
formula in the function field can be edited arbitrarily.

tives up to ninth order in Figure 5.1. The formulas for the preset functions cannot be
edited. The speed of computation is so great that the approximating polynomial reacts
to moving the computation point with the mouse virtually in real time.

The figure shows the ninth approximation for the Gauss function, which can be selected with the choice boxes above. In the number fields we now have the coefficients
of the Taylor series. They only differ from the values of the derivatives via the factor
$\frac{1}{n!}$ for the order n.

In the following simulation of Figure 5.3 a *parser* is used to evaluate functions that
can be edited. Using this simulation you can study the Taylor expansion for arbitrary

functions albeit at a slower speed of computation. Here the highest order is limited to 7.

The Taylor approximation of the red function is shown in blue and the deviation is plotted in green. Figure 5.3 shows a Gaussian function $y = f(x) = e^{-\frac{x^2}{b^2}}$ with the third approximation in the vicinity of the computation point, which is drawn in magenta and can be pulled with the mouse along the function. Using the keys $+1$ and -1, the approximation order can be increased and decreased.

This simulation allows for many possible experiments. In the selection field for functions, a number of standard functions can be selected (*sine, exponential function, power function, Gaussian, hyperbolic functions,* $\sin(x^2)$). They contain up to three parameters and can be edited. You can also enter an arbitrary analytical function for the computation.

Using a parser for the evaluation of the editable function slows down the computation considerably. Depending on the configuration of your computer, it can take up to a few minutes until the result for the seventh approximation appears.

After opening the simulation you first call a function from the selection list for which initially the third approximation is calculated for a computation point of $x = 0.5$. You can then move the computation point and change parameters, and the result is still shown practically in real time for the third approximation. The description pages of the simulation contain further details and suggestions for experiments that can be done.

5.5 Graphical presentation of functions

In Chapter 6 we will show interactive simulations that visualize functions in the plane, curves in space, surfaces and time-dependent surfaces. At this point we will give a short overview of the basic possibilities for visualizing functions.

5.5.1 Functions of one to three variables

Functions of one variable

Functions $y = f(x)$ are represented graphically in a two-dimensional system of coordinates, on which the *independent variable* is usually shown on the abscissa and the *dependent variable* $y = f(x)$ on the ordinate. An interval on the x-axis is mapped to an interval on the y-axis. The mapping is only unique if there is only one function value $y_1 = f(x_1)$ for a certain value x_1 of the independent variable. If one wants, for example, to show a circle, one needs to use two unique functions y_1 and y_2 for the parts of the circle above and below the abscissa:

$$x^2 + y^2 = r^2 \rightarrow y_1 = +\sqrt{r^2 - x^2}; \; y_2 = -\sqrt{r^2 - x^2}.$$

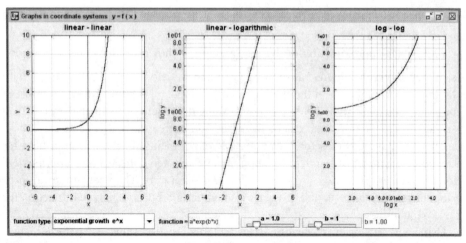

Figure 5.4. Simulation. Choice of different functions in linear and logarithmic coordinates. The picture shows the exponential function. *a* and *b* are adjustable parameters.

$y = f(x)$ with linear or logarithmic scaling of axes

The special character of a function can be underlined if one uses logarithmic scaling on one or both axes. With single-logarithmic presentation, exponential functions appear as lines and with double-logarithmic presentation, powers as lines. In addition, one can highlight regions of interest on the abscissa or ordinate using logarithmic stretching or compression. One also uses logarithmic scaling if one or both of the variables cover a very large range of values.

Figure 5.4 provides a simulation showing a number of preset functions next to each other in linear–linear, linear–logarithmic and double-logarithmic scales. The formula field is editable, which allows you to study arbitrary functions in comparison.

Further details and suggestions for experiments are given in the description pages.

Parameter representation of curves in plane and space

To show curves in the plane, that are non-unique with respect to the mapping from x to y, one uses the *parameter representation*, where both x and y are unique functions of a third independent variable, namely the *parameter p*.

$$x = f(p); \quad y = g(p)$$

$$p_1 \leq p \leq p_2.$$

For the circle around the origin with radius r this is, for example:

$$x = r \cos \varphi; \quad y = r \sin \varphi$$

$$\rightarrow x^2 + y^2 = r^2(\sin^2 \varphi + \cos^2 \varphi) = r^2 \cdot 1 = r^2$$

where the parameter ϕ is the angle between the radius vector to the computation point and the x-axis.

Using the parameter representation one can represent functions in the plane, which cover the coordinate ranges x and y multiple times, such as spirals.

Extending the parameter representation to the three coordinates of space one can visualize *space curves* in this way:

$$x = f(p); \quad y = g(p); \quad z = h(p).$$

Thus the number line is mapped to a line in the plane or in space.

Unique surfaces in space

With $z = f(x, y)$ one can represent surfaces in a three-dimensional space. Thus the surface is a mapping of the xy-plane with a height profile that depends on x and y. On paper one can only show two-dimensional projections of this surface. The technique of simulations extends this view quite dramatically, since it enables you to change the projections interactively or automatically, such that the impression of three dimensions being present is received; we will use this approach intensively in the following subsection.

Parameter representation of surfaces in three-dimensional space

Using a parameter representation with two parameters one can represent surfaces in spaces that are not unique with respect to a plane of reference, for example the surface of a sphere or a torus with respect to the xy-plane. In these cases, one needs $f_1(x, y)$ and $f_2(x, y)$. In parameter representation one writes:

$$x = f(p, q); \quad y = g(p, q); \quad z = h(p, q).$$

Thus the two number lines p and q are mapped to a surface that lies in space.

Functions of three variables

A density distribution, for example of charge or mass in space, is described using a function D of the three spatial coordinates x, y and z, i.e. $D(x, y, z)$. How can such functions of three variables be visualized? One obviously needs another variable beyond the three space coordinates.

A qualitative option consists of assigning to a regular space grid of points a color coding for $D(x, y, z)$ and choosing the density of points in such a way that the space stays "transparent". The grid is then projected on a surface, and changing the projection as a function of time again increases the spatial impression.

A second option is to use surfaces in space on which $D(x, y, z)$ has a constant value. One can then stagger semi-transparent surfaces inside each other, or the constant value of each surface can be changed as function of time. In the moving projection both possibilities yield a quantitative picture. In the first case one uses the opacity, and in the second case the time, as the additional variable.

5.5.2 Functions of four variables: World line in the theory of relativity

A physical event such as the ticking of the watch on my wrist takes place in a three- World dimensional space (x, y, z) and, because it depends on the time t, it can be considered as a four-dimensional function E:

$$E = f(x, y, z, t)$$

In a simulation this can be represented, for example, by calculating for a cohort of three-dimensional functions $E(x, y, z, t_i)$ for a number of discrete points t_i in time, two-dimensional projections, and displaying these one after the other. In general this method has, of course, limited applicability. It is relatively simple, when dealing with a chain of events that have a lower dimensionality, for example in the case of a propagating surface in space during an explosion. In general one will restrict this method to a lower dimensional projection.

This is especially the case when describing phenomena in the special theory of relativity. In this theory, the time t joins the three spatial dimensions as the "fourth dimension". In order for this variable to have dimensions of length, one usually normalizes t via multiplication with the velocity of light $c = 3 \times 10^8$ m/sec:

$$E = f(x, y, z, ct) \quad \text{or} \quad E = f(x_1, x_2, x_3, x_4).$$

A four-dimensional chain of events – for example an exploding supernova – can only be visualized with difficulty as a *whole*. To capture this phenomenon in its entirety one would like to imagine the whole explosion in a single moment. Indeed, *Homer* and the pre-socratic philosophers speculated around 500 BC about the *god-like* possibility of recognizing space and time as past, present and future, as a *unit*. Around the year 520, *Boethius* formulated such thoughts in his work *Comfort of Philosophy*.

One circumvents this problem by doing away with two space dimensions for visualizations in the theory of relativity and plotting the chain of events on a plane diagram, for example with the time dimension ct on the ordinate and, on the abscissa, the space dimensions, x, in which the event takes place. The event chain of a body that is moving in the x-direction is then called the *world line*.

This will be visualized with an example in Figure 5.3, where a point object is moving with constant acceleration in the x-direction. The limiting velocity of light, i.e. the *world line* of a light flash $x = ct$ is shown in red; in black, the event chain that

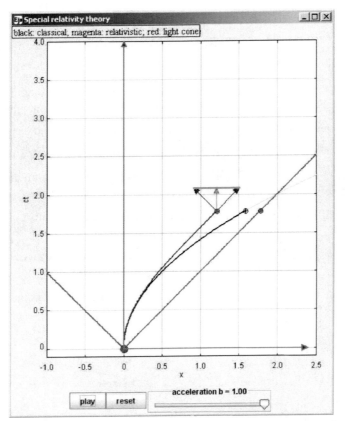

Figure 5.5. Simulation. World line (blue) of a point object (magenta) that is accelerated uniformly in one dimension. The black line shows the classically "possible" orbit, the blue line the orbit that can be realized according to the special theory of relativity. The red lines delimit the light cone of a light signal sent at the same time. The arrows originating from the object delimit its light cone. The constant acceleration can be adjusted with the slider.

would be possible according to classical mechanics, for which arbitrarily large velocities could be achieved; and in magenta, the actually possible event chain according to the theory of relativity, for which the velocity will approach the velocity of light but will not exceed it. The arrows show the respective *light cones* in which all events caused by the object happen – all as seen from an observer at the origin.

In this simulation the movement of the objects is shown as a function of time. Further details are given in the description of the interactive simulation.

In articles on the *special theory of relativity* the time is normally plotted on the ordinate and the space on the abscissa, such that the light cone opens up to the top. The classical acceleration parabola $x = \frac{b}{2}t^2$ is then open to the right.

5.5.3 General properties of functions $y = f(x)$

In the following we define important properties of a function of one variable $y = f(x)$ on its domain of definition D. A function $f(x)$ is:

bounded,	if in the interval of definition D there is a maximum value (supremum) and a minimum value (infimum);
one-sided continuous	if $f(x)$ continues smoothly in one direction;
continuous in x_0,	if $f(x)$ continues smoothly in both directions;
continuous in the domain of definition D,	if $f(x)$ is continuous at all points in the domain of definition, i.e. there is no jump;
one-sided differentiable in x_0,	if $f(x)$ has at x_0 a unique derivative in one direction;
differentiable in x_0,	if $f(x)$ has at x_0 the same derivative in both directions;
differentiable in D,	if $f(x)$ is differentiable in every point of D.

Corresponding examples are shown in Figure 5.6:

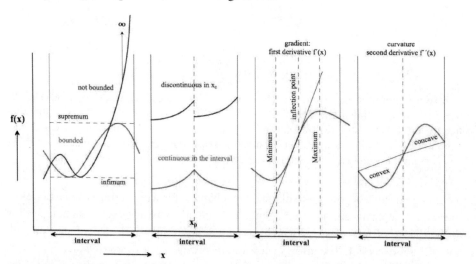

Figure 5.6. Properties of function graphs $f(x)$ in an interval of the variable x. In the example on curvature the notations concave and convex are defined for a finite interval.

The slope of the curve at a point is characterized by the direction of the tangent in the point and thus by the differential quotient $f'(x)$. At the maximum and minimum, i.e. at the extrema, the tangent is parallel to the x-axis and the tangent of the slope angle is zero. At the turning point the slope attains its largest or smallest value with respect to its vicinity; in the example we have a turning point with a positive slope.

The curvature of the graph of a function on an interval is defined as follows:

Concave function graph (negative curvature): all points on the graph lie above the chord connecting the end points of the graph on the interval; the slope increases with increasing x.

Convex function graph (positive curvature): all points on the graph lie below the chord connecting the end points of the graph on the interval; the slope decreases with increasing x.

The curvature at a point (x, y), considered as an infinitesimal interval, is obtained as a limit for the vanishing width of the interval. It describes the change of the slope and therefore is equal to the second derivative $f''(x)$. Thus this quantity, as well as the curvature, is a local quantity and, in addition, a function of x itself for those functions that can be differentiated twice. At the turning point the sign of the curvature changes; the curvature and thus the second derivative $f''(x)$ vanishes at the turning point.

The red curve in the second example has a kink at x_0 where no unique slope is defined, but only a right-sided and left-sided derivative exist. It is therefore only one-sided differentiable; its derivative is not continuous at x_0. The blue curve in the first example diverges at the end of the interval; it is not bounded and has no supremum.

5.5.4 Exotic functions

Functions can be of many different types. One of these, which is often given in text-books and is both simple and exotic, gives food for thought, but is very well defined; it is given by

$$f(x) = \begin{cases} 1 & \text{for } x \text{ irrational,} \\ 0 & \text{for } x \text{ rational} \end{cases}$$

domain of definition $0 \geq x \geq 1$.

This function can obviously not be visualized graphically, since there are infinitely many rational and irrational numbers in the domain of definition, such that the values of 0 and 1 are nested indissolubly on the ordinate. This function is not continuous at any point and cannot be differentiated anywhere.

A graphically attractive exotic function is the fractal *Koch curve*, which is obtained Koch curve etc. as the limit of a combination of triangular lines. This function is, in contrast to the above, continuous, but does not have a well defined slope and thus no derivative.

The functions that are important for physics are, however, mostly well-behaved, with the exception of a few points. The following section demonstrates a few typical properties.

5.6   The limiting process for obtaining the differential quotient

After these preliminary discussions we want to visualize the limiting process involved in differentiation in a simulation for the sine function. Figure 5.7 shows the sine function $y = \sin x$ over somewhat more than a full period. The first (analytical) derivative, the cosine function $d(\sin x)/dx = \cos x$, is drawn in yellow. A blue point, at which the limiting process will be observed, can be adjusted with the slider on the plot of the sine function. The large red point can also be adjusted along the sine curve. The line connecting these two points is extended in green.

The red and blue arrows show the difference of the ordinates (Δy) and abscissae (Δx) between the movable red and the fixed blue point. The magenta point indicates the value of the difference quotient $\Delta y/\Delta x$. If you pull the red point to the blue point

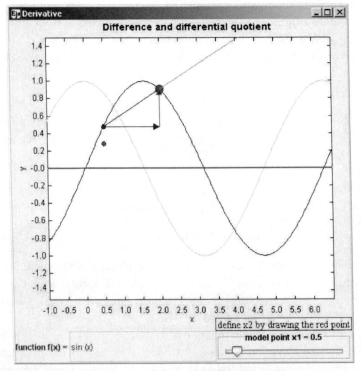

Figure 5.7. Simulation. Visualization of how the difference quotient approaches the differential quotient in the limit of $\Delta x \rightarrow 0$ for the example of the sine function (black) and its first derivative (yellow). The position of the computation point in blue can be changed with the slider and the red point can be moved with the mouse. The small magenta point indicates the value of the respective difference quotient. With decreasing width of the abscissa interval it approaches the analytical differential quotient.

the line connecting them becomes the tangent and the point for the difference quotient moves to the curve for the first derivative. This is the limiting process $\Delta x \to 0$ of the difference quotient. You can reconstruct the curve of the first derivative by moving the blue computation point along the sine curve. In the description pages you will find hints for further useful experiments.

The difference quotient obviously does not change, if the curve of the function that is drawn symmetrically to the red colored x-axis is moved up or down by a constant value c, the same applies to the differential quotient. This corresponds to the rule that the derivative of a constant vanishes. All functions that are different only by a constant value in the y-direction have the same derivative:

$$\frac{d}{dx}(f(x) + c)) = \frac{d}{dx}f(x).$$

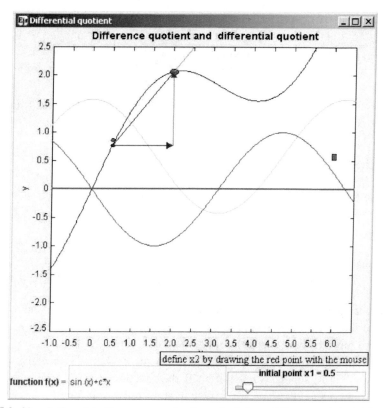

Figure 5.8. Simulation. Limiting process for the calculation of the second derivative (blue) for a sine function (black) that has been supplemented by a linear term. The computation point and the width of the interval can be adjusted using the slider and pulling the red point with the mouse and the linear term can be changed by pulling the rectangular purple marker. The first derivative drawn in yellow is then moved in the y-direction.

Applying the same line of thought to the determination of the second derivative (the figure shown above is also valid if one interprets the black curve as the first derivative and the yellow as the second derivative), it follows:

$$\frac{d^2}{dx^2}(f(x) + c_1 + c_2 x) = \frac{d^2}{dx^2} f(x).$$

The second derivative (the curvature of the original function) is identical for all functions that only differ by a constant c_1 and a linear term $c_2 x$.

This is visualized via the simulation in Figure 5.8, where the second derivative $-\sin x$ is plotted in addition. A purple rectangle is also present, which can be pulled to add a linear term $c_2 x$ to the sine function. This results in the first derivative being shifted by the value c_2 in the y-direction. The magenta colored point of the difference quotient is again led to the curve of the first derivative via the limiting process. The second derivative is not affected by changing c_2.

The second derivative characterizes a function up to two constants, which are *initial values* of the function, namely the value of the function itself and the first derivative at a point chosen as $x = 0$, without loss of generality. From the cohort of all functions that have the same second derivative, only the initial values determine a unique function.

This train of thought can also be applied to higher derivatives. The nth derivative characterizes a cohort of curves with n parameters.

5.7 Derivatives and differential equations

For the sine function we have a simple relationship between the function and its second derivative; it is equal to the *negative* sine function. The same relationship applies to the cosine function.

$$y = \sin(x) \to y' = \cos(x) \to y'' = -\sin(x) \Rightarrow y'' = -y$$
$$y = \cos(x) \to y' = -\sin(x) \to y'' = -\cos(x) \Rightarrow y'' = -y.$$

For these trigonometric functions, the differential equation expresses the fact that the absolute value of the derivative is equal to the function value *having the opposite sign*. What does this mean in concrete terms?

If the function value y is positive and large, the curvature is negative and large, leading quickly to smaller values of y. If the function value is negative and has a large absolute value, the large positive curvature quickly leads to a larger value. If the function value is small, the curvature is also small and therefore an increase or decrease continues nearly linearly as at a turning point.

The negative relationship between the function and its curvature thus leads to oscillating behavior. You are encouraged to confirm these statements in the previous two figures.

The fact that both trigonometric functions $\sin x$ and $\cos x$ satisfy the same differential equations shows their close relationship as oscillating functions. It follows immediately that the sum of sine and cosine functions satisfies the same differential equation. (Also check that this sum is, according to the addition rules for trigonometric functions, identical to a *phase-shifted function*.) As a second example we consider the *exponential function* for both positive and negative exponents:

$$y = e^x \rightarrow y' = e^x; \qquad y'' = e^x \rightarrow y' = y \quad \text{and} \quad y'' = y$$
$$y = e^{-x} \rightarrow y' = -e^{-x}; \quad y'' = e^x \rightarrow y' = -y \quad \text{and} \quad y'' = y.$$

Now the second derivative also has the same sign as the original function. What do these relationships mean in concrete terms?

The curvature is equal to the function value. The larger the function value, the larger the curvature. Any curvature already present increases with increasing y. If the slope (first derivative) has the same sign as the function, the function will grow faster and faster beyond any boundaries – it diverges. If the slope has the opposite sign to the function, the function decreases faster and faster to zero; it converges to 0. The differential equation $y'' = y$ describes both behaviors.

As shown for the trigonometric functions, the differential equation is then also valid for the sum of two exponential functions. If one takes exponents with different signs for the two functions, the hyperbolic functions are covered:

$$\sinh(x) = \frac{e^x - e^{-x}}{2}; \quad \cosh(x) = \frac{e^x + e^{-x}}{2}.$$

Thus the differential equation $y'' = y$ describes the exponential and hyperbolic functions and this common property shows their close relationship.

Differential equations describe the local, *internal* structures of function, their *character*, and they are the "generators" of cohorts of related functions.

5.8 Phase space diagrams

All variables of a system constitute its *phase space*. A selection of a few variables is referred to as a *phase space projection*. For a differential equation $y' = y'(y, x)$, $y(x)$, $y'(x)$ and $y'(y)$ are three meaningful projections of the phase space.

The general characteristics *divergent/convergent/oscillating* of a differential equation can be visualized well in a diagram that shows, in addition to the function $y(x)$ and its derivative $y'(x)$, the projection $y'(y)$.

In Figure 5.9 the phase space projection for the system $y(x) = \sin(nx)$ with the differential equation $y' = dy/dx = n \cos nx = n\sqrt{1 - \sin^2 nx} = n\sqrt{1 - y^2}$ is

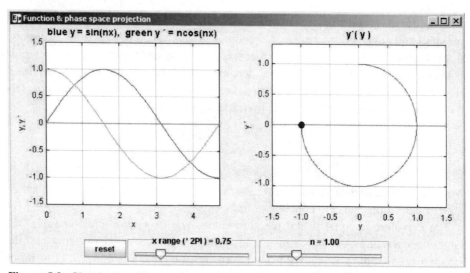

Figure 5.9. Simulation. Phase space projections for $y = \sin(nx)$ ($n = 1$ in the figure). The left window shows $y(x)$ in blue and $y'(x)$ in green. The zero line is marked in magenta. The right window shows $y'(y)$. The parameter x *range* determines the size of the interval, the parameter n the number of periods in the interval. The blue point in the phase space is the end point of the interval.

shown in the right-hand window. The adjustable constant n determines the number of periods in the interval $0 \le x \le 2\pi$.

For the case of the trigonometric function, $y'(y)$ is for $n = 1$ a circle that is transversed periodically; for $n < 1$ the curve becomes an ellipse because of the factor n, and is not closed (why?). For $n > 1$ this ellipse is transversed multiple times.

In this case, the differential equation is particularly simple. More complex differential equations of order n define families of more involved functions. One can, however, always differentiate between solutions that converge, diverge or oscillate with increasing variables, and the phase space projections $y^{(n)}(y)$ make this difference particularly apparent.

Later we will visualize solutions of differential equations in more detail.

5.9   Antiderivatives

5.9.1   Definition of the antiderivative via its differential equation

The first derivative, the differential quotient, describes the change of a given function $y = f(x)$ in its dependence from the variable x. We can now ask the converse question: Is there a function $F(x)$ whose change is described by $f(x)$, and what properties does this function have? If such a function exists, it is called the *antiderivative* of $f(x)$

or its *indefinite integral*. It is described by a very simple differential equation:

$$F'(x) = f(x), \quad f \text{ given, } F \text{ wanted}$$

$$F(x) = \text{Integral of } f(x) = \int f(x)dx$$

$$\rightarrow \frac{d}{dx} \int f(x)dx = f(x).$$

The integral sign serves as a reminder that the calculation proceeds via a summation and the notation $f(x)dx$ reminds us that a limiting process takes place for the calculation for which the variable interval becomes infinitesimally small, that is, $\Delta x \rightarrow dx$; we will visualize this shortly.

This differential equation obviously defines a whole cohort of functions, which can differ by a constant value, because the derivative (change) of a constant vanishes. Thus the indefinite integral of a given function is known up to a constant.

$$\frac{d}{dx}(F(x) + C) = \frac{d}{dx}F(x) = f(x).$$

If the differential equation has a meaningful solution, i.e. if the function is *integrable*, the indefinite integral is analogous to the differential quotient of a function, which describes, up to a constant, a *local property* of the integrated function $f(x)$.

5.9.2 Definite integral and initial value

What is the meaning of the *integration constant*? As long as we do not decide on the range of the variable x it is simply an arbitrary number.

If, however, we start at a certain *initial value* x_1 and take into account that $f(x)$ is the *change* $F'(x)$ of the antiderivative, then the antiderivative describes the process of the changes in $F(x)$ given by $f(x)$ from the variable value x_1 onwards.

We now show this in a simple example from physics: we assume that $f(t)$ is the time-dependent velocity, $v(t)$, of an object. The result of this time-dependent velocity, which can also have negative values, is the distance traveled $F(t)$ i.e. $x(t)$. Thus $v(t)$ determines the distance from the initial point as a function of time.

The constant C is the *initial value* $F(x_1)$ of the integral for the variable x_1, in our example the position from which we start.

Provided the range of the variable is open, i.e. $x > x_1$, the *definite integral* defined in this way is a function of the variable x.

If we are interested in the behavior of the antiderivative in a *closed interval* $x_1 \leq x \leq x_2$, the definite integral becomes a *fixed value*. The value at the end of the integration range is the result of the initial value and of all changes until the final value of x, and is given by the antiderivative $F(x_2)$. The change within the interval results from the difference to the initial value. Calculating this difference also gets rid of the unknown integration constant, if we repeat the same line of thought for the initial and final value with an arbitrary initial value outside the interval:

$$\int_{x_1}^{x_2} f(x)dx = F(x_2) + C - (F(x_1) + C) = F(x_2) - F(x_1).$$

This relationship is known as main theorem of differential and integral calculus.

Thus, in order to calculate a definite integral we "only" need to know its antiderivative. To determine the antiderivative for an arbitrary function $f(x)$ is, in general, not as easily possible as for the derivative. Basic functions can be easily integrated by inverting the well known relations for their derivatives; for many complicated functions there are tables. There are also quite a few useful general rules, which can help to find the antiderivative, for example "integration by parts". But there is, unfortunately, no rule that *always* succeeds.

Therefore, numerical methods play an especially important role for integration, as we will discuss later.

5.9.3 Integral as limit of a sum

In analogy to calculating the partial sums of a series, in an xy-plot of the function one can define the integral as the *surface measure* of the function value in an interval of the variable. It is obvious that one cannot simply calculate a sum of function values, since their number would be infinitely large. The factor to be used is analogous to the index difference for series and is equal to the width of the interval. If one multiplies this factor with a suitably chosen function value we obtain a measure for the surface under the function in the interval.

Since functions change in general when the variable changes, choosing an arbitrary function value from the interval (for example at the beginning, in the middle or at the end) can only yield an approximation. In this case, one decomposes a larger interval $[x_1; x_2]$ into n intervals chosen equal for expedience of width $n = (x_2 - x_1)/\Delta x$ and sums over the approximate measures of the sub-intervals. Then the integral is defined as the limit of this sum for a vanishingly small sub-interval.

Measure of the sub-interval Δx: $\quad f(x_i)\Delta x; \; x_i \text{ in } \Delta x$

Total measure of the region $x_2 > x > x_1$: $\quad \sum_{i=1}^{n} f(x_i)\Delta x$

Integral: $\quad \int_{x_1}^{x_2} f(x)dx = \lim_{\Delta x \to 0} \sum_{i=1}^{n} f(x_i)\Delta x.$

The definite integral provides the area between the function $f(x)$ and the x-axis in the region of integration.

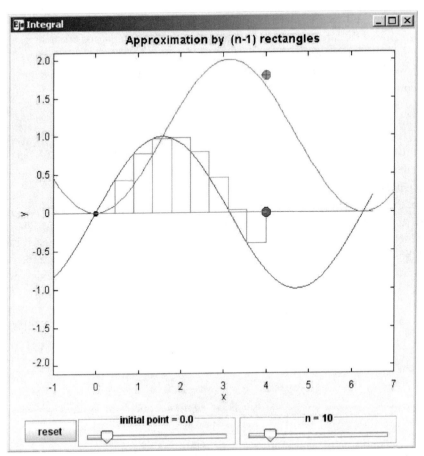

Figure 5.10. Simulation. Limiting processes for the integration using the step function approximation (green), shown for the example of the sine function (blue). For each interval, the initial value is assumed to be valid. The red curve is the antiderivative, the point filled in green indicates the approximation for the definite integral in the integration region whose initial and final point can be adjusted. The number of intervals n ($n = 10$ in the figure) can be adjusted with the slider.

The limiting process is shown in the interactive simulation of Figure 5.10

The sine function to be integrated is drawn in blue, while the analytical *integral* function is drawn in red. The small blue point, which can be moved with the slider, indicates the initial point for the integration and thus, at the same time, the zero point of the formal integral. The thick end point in magenta can be adjusted with the mouse. The second slide determines the number of sub-intervals .

The green rectangles represent the contribution for the individual interval, if the initial value of the function in the interval is assumed to be constant for the whole interval. The sum of the contributions for all intervals yields the large green point. With decreasing width of the intervals, it approaches the analytically calculated integral. For a sufficiently large number of intervals this value runs along the integral curve when pulling the end point.

You will find further instructions for experiments in the description pages of the simulation.

5.9.4 The definition of the Riemann integral

We still require a criterion to enable us to decide whether a function can be *integrated* at all in a given region. In the *classical sense* this is provided by the integral definition of *Riemann*.

For this purpose we define for the intervals given by $x_0 < \cdots < x_i < \cdots < x_n$ with interval widths $x_i - x_{i-1} = \Delta x_i$ two sums, namely the *upper sum* and the *lower sum*, of which the first one uses the largest function value, the supremum, in each interval and the second one uses the smallest function value, the infinum, in each interval. If both sums converge to the same value for $n \to \infty$, the one from above the other one from below, the function is considered as *integrable* in the *Riemannian sense*

First measure for sub-interval $\Delta_i x$: $\Delta_i x \cdot$ *supremum* of $f(x)$ in $(\Delta_i x)$

Second measure for sub-interval $\Delta_i x$: $\Delta_i x \cdot$ *infimum* of $f(x)$ in $(\Delta_i x)$

First sum measure for region $x_2 > x > x_1$:

$$\sum_{i=1}^{n} \Delta_i x \cdot \textit{supremum of } f(x) \textit{ in } (\Delta_i x)$$

Second sum measure for region $x_2 > x > x_1$:

$$\sum_{i=1}^{n} \Delta_i x \cdot \textit{infimum of } f(x) \textit{ in } (\Delta_i x)$$

the Riemann Integral $\displaystyle\int_{x_1}^{x_2} f(x)dx$ exists, if

$$\lim_{n \to \infty} \sum_{i=1}^{n} \Delta_i x \cdot \textit{supremum of } f(x) \textit{ in } (\Delta_i x)$$

$$\overset{!}{=} \lim_{n \to \infty} \sum_{i=1}^{n} \Delta_i x \cdot \textit{infimum of } f(x) \textit{ in } (\Delta_i x).$$

In the following interactive simulation shown in Figure 5.11, the construction of the Riemannian sums is demonstrated using the example of the sine function. In the left window the upper sum (supremum) is used and in the right window the lower sum (infinum). The width of all intervals is the same. The analytical integral is shown in

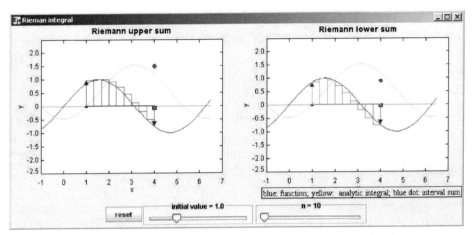

Figure 5.11. Simulation. Limiting process for the Riemann integral for the example of the sine function (black); the antiderivative is yellow. Integration region and number of intervals can be adjusted, 10 intervals in the figure. For the upper sum the highest value is used and for the lower sum the smallest value is used in each interval. The rectangular markers indicate the approximations for supremum (left window) and infimum (right window).

yellow. The initial and final x-values can again be adjusted as well as the number of intervals. With increasing resolution both sums tend to the same value.

The initial x-value can again be adjusted with a slider and the final x-value (magenta) can be pulled with the mouse. The number n of sub-intervals in the integration region is adjusted with the second slider. The analytical determined integral is indicated in yellow. Its initial value is given by the initial ordinate of the integration region. The point that is surrounded by a square shows the sum of approximating intervals.

If it is known that a function is *Riemann*-integrable, then any sum that uses, as a measure, any value of the function in the sub-intervals, converges against the integral. Thus one has a lot of freedom in the choice of numerical integration method. You are urged to compare the last two figures. The step-function approximation is neither equal to the approximation with the supremum nor to that via the infinum, but converges to the same limit.

As an example for a function that *cannot* be integrated in the Riemannian sense, the *exotic* function mentioned above, can be considered:

$$f(x) = \begin{cases} 1 & \text{for } x \text{ irrational,} \\ 0 & \text{for } x \text{ rational} \end{cases}$$

domain of definition $0 \geq x \geq 1$.

In its domain of definition it has obviously an upper sum 1 and a lower sum 0, since there are both rational and irrational numbers in every interval of an arbitrarily small length $\Delta x > 0$, and thus there exist function values of 0 and 1. Thus the upper sum and the lower sum converge, but not to the same value, and therefore the function is not *Riemann*-integrable.

5.9.5 Lebesgue integral

The previous statement is not really satisfactory. The number of rational numbers is much smaller than that of the irrational ones, and therefore the function $f(x)$ has the value 1 for nearly all values of x. Therefore, the integral of this function should be close to 1.

This question can be more easily answered with the alternative notion of the *Lebesgue*-integral. For this approach one subdivides the integration region in stripes *parallel* to the x-axis and asks for the limit of the sum over these intervals, each interval contributing the product of the function value in the interval and the corresponding *Lebesgue*-measure of the interval on the ordinate:

$$\mu(\Delta y) = \text{Measure of all } x\text{-values, whose } f(x) \text{ lie in } \Delta y.$$

In the *exotic* example the top stripe has the function value 1 and the measure of its variable interval is (for the moment approximately) 1, since nearly all numbers are irrational. The lowest strip has the function value 0, independent of the measure for the variable interval.

The exotic function is therefore *Lebesgue*-integrable and the result is 1.

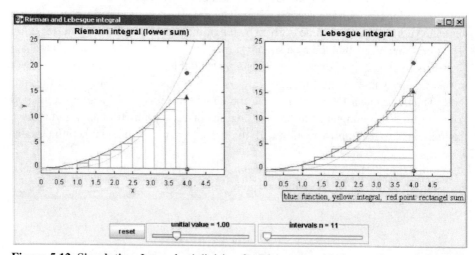

Figure 5.12. Simulation. Interval subdivision for Riemann and Lebesgue integral; the function is shown in blue, the antiderivative in yellow and the red points indicate the approximation for the chosen number of points n. The integration region can be adjusted. For the Lebesgue integral the correct measure for the limit was already used.

The advantage of the integral definition of Lebesgue is that, when using it, the integral notion can go beyond the domain of numbers to *sets* in general, if these sets can be decomposed into subsets, which can each be *measured* in the sense of a finite area. The following holds: a function that is Riemann-integrable is also Lebesgue-integrable but the converse is not always true. Thus the Lebesgue integral is the more general notion.

In the simulation of Figure 5.12 we visualize the integration of a parabola on the left-hand side using Riemann's approach and on the right-hand side with Lebesgue's approach. For the Lebesgue integral, the interval measure was calculated in such a way that the measure is exact irrespective of the width of the interval.

5.9.6 Rules for the analytical integration

As for derivatives, there are a number of important and general rules (we drop the integration constant in the following for clarity).

$$\int C dt = C \int dt = Ct \quad \text{constant } C$$

with $g = g(t)$ and $h = h(t)$

$$\int (g(t) + h(t))dt = \int g(t)dt + \int h(t)dt \quad \text{Additivity}$$

$$\int g dh = gh - \int h dg \quad \text{Integration by parts}$$

$$\int f(t)dt = \int f(g(x))g'(x)dx$$

Introduction of a new variable x via $t = g(x)$.

For the particularly useful rules of partial integration and substitution of a new variable, it is important to find functions that can be easily integrated, as for example *the exponential function, powers of* x and *the trigonometric functions.*

The following formulas for basic functions without the integration constant follow very easily from the formulas given above for the first derivatives and therefore we only list those with the greatest practical importance:

$$\int C dt = Ct; \quad \int t^n dt = \frac{t^{n+1}}{n+1}; \quad \int e^t dt = e^t;$$

$$\int a^t dt = \int e^{t \ln a} = \frac{a^t}{\ln a}; \quad \int \sin t dt = -\cos t;$$

$$\int \frac{1}{t} dt = \ln t; \quad \int \cos t dt = \sin t.$$

The analytical integration of complex functions that can be integrated in principle is, as a rule, more tedious than the always easily achievable differentiation. Therefore there exist voluminous collections of integrals in the corresponding text books, manuals and on the internet. Numerical computer programs such as *Mathematica* also have a wide range of formal integrals built in, which one can access as formulas if one enters the function to be integrated.

It is obvious that numerical integration methods play a very important role, since it does not matter for their application whether an integral of the function to be integrated is known analytically or not, and since one can even integrate functions that are only known as discrete measured values f_i.

5.9.7 Numerical integration methods

Integrals often have to be calculated numerically, if it is not possible to determine the antiderivative analytically. In this case the sums obtained using step functions converge only relatively slowly when decreasing the interval widths; one would therefore have to subdivide the integration region into numerous sub-intervals to achieve a high level of accuracy.

Therefore, approximations of the function $f(x)$ other than by step functions can be used in order to reach convergence faster. An obvious approximation when looking at Figure 5.12 consists of not taking the value $f(x_i)$ at the beginning of the interval as constant for the interval (*step-function rectangle approximation*), but to use the mean value between the initial and final values $\frac{1}{2}[f(x_i) + f(x_i + 1)]$. This corresponds to a *trapezoidal approximation*, where one adds to the staircase the triangle leading to the next function value; the curve is now approximated via the initial value in the interval and the secant connecting the final and initial value with the slope $\frac{y_{i+1}-y_i}{\Delta x}$.

The approximation of the function becomes even more accurate if one uses a parabola (*Simpson's/Kepler's method*) that is fixed via three consecutive function values. This now also takes the curvature (second derivative) in each interval into account approximately. Thus those regions of the function that possess, like a parabola, no turning points in the respective sub-intervals (x_i, x_{i+1}), are approximated well. One can continue is this manner if one uses polynomials of third or fourth degree, which then also allow for the representation of turning points. However, one then needs to use more and more intermediate points in each sub-interval. Therefore, this approach is usually restricted to the parabola, where a sufficiently small interval is chosen.

All these methods have the advantage that the approximation of the function in terms of constants, secants and parabolas can be quite easily integrated in the interval.

$$\text{Rectangle approximation: } y = y_i \rightarrow \int_{x_i}^{x_i+\Delta_i x} y\,dx \approx \int_{x_1}^{x_1+\Delta_i x} y_1\,dx = \Delta_i x \cdot y_i$$

Trapezoidal approximation: $y = y_i + \dfrac{y_{i+1} - y_i}{\Delta_i x}(x - x_i) \rightarrow$

$$\int_{x_i}^{x_i + \Delta_i x} y\, dx \approx \Delta_i x y_i + \frac{y_{i+1} - y_i}{\Delta_i x}\left[\frac{(x_i + \Delta_i x)^2 - x_i^2}{2} - x_i(x_i + \Delta_i x - x_i)\right]$$

$$= \Delta_i x y_i + \frac{\Delta_i x}{2}(y_{i+1} - y_i) = \frac{\Delta_i x}{2}(y_i + y_{i+1})$$

Parabolic approximation: $\displaystyle\int_{x_i}^{x_i + \Delta_i x} y\, dx \approx \int_{x_i}^{x_i + 2\Delta_i x} (ax^2 + bx + c)\,dx$

$$= \frac{\Delta_i x}{3}(y_i + 4y_{i+1} + y_{i+2}).$$

The simulation in Figure 5.13 compares the three methods for two adjacent sub-intervals. As an example we again consider the sine function (blue) with its analytical integral (red). Initial and end point of the integration region can be changed. The sum of both sub-intervals is shown as a green point. The simulation shows the great superiority of the parabolic approximation, whose result agrees with the red curve even for a coarse subdivision of the interval.

Calculating the parameters of a parabola that goes through three points can be tedious, but this is only necessary if, as for this simulation, the osculating curves are calculated. The following steps are required for the calculation in each sub-interval $x_i = x_1, \frac{1}{2}(x_i + x_{i+1}) = x_2, x_{i+1} = x_3$:

coordinates in the interval x_1, y_1 x_2, y_2 x_3, y_3

with $x_2 - x_1 = \Delta_i x/2$; $x_3 - x_2 = \Delta_i x/2$; $\Delta_i x = x_3 - x_1$

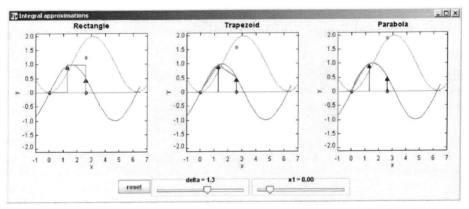

Figure 5.13. Simulation. Step-function, trapezoidal and parabolic approximations for the numerical integration of the sine function (blue) with two sub-intervals. When reducing the size of the intervals one can compare the convergence of these approximations. The closer the numerical value (green) point is to the known analytical curve, the better the approximation method performs.

general parabola $y = ax^2 + bx + c$

the parameters a, b, c are determined from the function values:

$$1 \rightarrow y_1 = ax_1^2 + bx_1 + c, \; 2 \rightarrow y_2 = ax_2^2 + bx_2 + c, \; 3 \rightarrow y_3 = ax_3^2 + bx_3 + c.$$

Solution for a, b, c yields

$$a = \frac{2}{\Delta x^2}(y_1 - 2y_2 + y_3)$$

$$b = \frac{2}{\Delta x}(y_2 - y_1) - (x_1 + x_2)a$$

$$= \frac{2}{\Delta x}(y_2 - y_1) - \frac{2}{\Delta x^2}(x_1 + x_2)(y_1 - 2y_2 + y_3)$$

$$c = y_1 - ax_1^2 - bx_1 = y_1 - x_1^2 \frac{2}{\Delta x^2}(y_1 - 2y_2 + y_3)$$

$$- x_1 \left[\frac{2}{\Delta x}(y_2 - y_1) - (x_1 + x_2)\frac{2}{\Delta x^2}(y_1 - 2y_2 + y_3) \right].$$

For the approximation to the integral over the sub-interval $\Delta_i x$ one obtains, using the parameters of the parabola and integrating a surprisingly simple formula, for which only the three function values and the width of the interval are required.

$$\text{parabolic approximation of } \int_{x_i}^{x_{i+1}} f(x)dx \approx \int_{x_i}^{x_{i+1}} (ax^2 + bx + c)dx$$

$$= \frac{\Delta x_i}{6}(y_i + 4y_{\Delta_i x \over 2} + y_{i+1}).$$

5.9.8 Error estimates for numerical integration

To get an idea of the accuracy of the different integration methods, we expand the function in a Taylor series and use, assuming the interval is sufficiently small, the first neglected term as an estimate for the error. To simplify the notation, we expand the function in a Taylor series around $x = 0$ up to the fifth order:

$$f(x) = y(x) = y(0) + y'(0)x + y''(0)\frac{x^2}{2!} + y'''(0)\frac{x^3}{3!}$$

$$+ y^{(4)}(0)\frac{x^4}{4!} + y^{(5)}(0)\frac{x^5}{5!}$$

$$1) \quad \int_0^{\Delta x} f(x)dx = y(0)\Delta x + \frac{y'(0)}{2}\Delta x^2 + \frac{y''(0)}{3 \cdot 2!}\Delta x^3$$

$$+ \frac{y'''(0)}{4 \cdot 3!}\Delta x^4 + \frac{y^{(4)}(0)}{5 \cdot 4!}\Delta x^5 + \frac{y^{(5)}(0)}{6 \cdot 5!}\Delta x^6$$

$$2) \int_{-\Delta x}^{0} f(x)dx = y(0)\Delta x - \frac{y'(0)}{2}\Delta x^2 + \frac{y''(0)}{3\cdot 2!}\Delta x^3$$

$$- \frac{y'''(0)}{4\cdot 3!}\Delta x^4 + \frac{y^{(4)}(0)}{5\cdot 4!}\Delta x^5 - \frac{y^{(5)}(0)}{6\cdot 5!}\Delta x^6$$

$$\int_{-\Delta x}^{\Delta x} f(x)dx = \int_{0}^{\Delta x} f(x)dx + \int_{-\Delta x}^{0} f(x)dx$$

$$= 2\left[y(0)\Delta x + \frac{y''(0)}{3!}\Delta x^3 + \frac{y^{(4)}(0)}{5!}\Delta x^5 \right]$$

$$\int_{-\Delta x/2}^{\Delta x/2} f(x)dx = 2\left[y(0)\frac{\Delta x}{2} + \frac{y''(0)}{3!}\left(\frac{\Delta x}{2}\right)^3 + \frac{y^{(4)}(0)}{5!}\left(\frac{\Delta x}{2}\right)^5 \right].$$

For the *step-function* method we use only the first term $y(0)$ in 1). The error for each interval is thus of the order Δx^2. If one wants to know the error for the entire integration region, one has to sum over the $L/\Delta x$ intervals. Thus the total error is proportional to Δx. Doubling the resolution leads to halving of the error or doubling of the accuracy.

For the *trapezoidal* method the first two terms are used in 1). The interval error then is proportional to Δx^3, thus the total error depends on Δx^2. Doubling the resolution leads to an improvement in the accuracy by a factor of 4.

For the parabola method we expand the function from the middle of the double interval once to the right and once to the left and the integral over the whole interval is the sum over both sub-intervals. The result then only contains odd powers of Δx. For

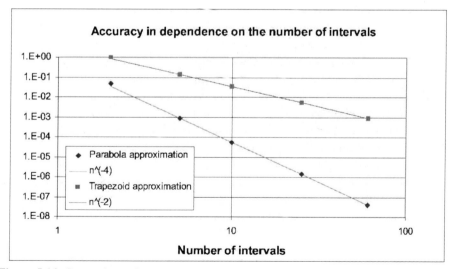

Figure 5.14. Comparison of accuracy achieved for the numerical integration using the trapezoidal and parabolic approximations as functions of the number of sub-intervals n. For 100 sub-intervals the parabolic approximation is at least *five* orders of magnitude more accurate.

the parabola we also take into account the curvature, i.e. y''. The error for each interval is then proportional to Δx^5, the total error is thus proportional to Δx^4; doubling the resolution leads to an increase in the accuracy by a factor of 16. In addition, the large factor $5! = 120$ contributes to a small error.

Important hint: the approximating parabola used for the integration is *not* identical with the third partial sum of the Taylor series. This one only agrees with the function at the computation point, while the approximating parabola used for the integration is *equal* to the function at all three points.

Figure 5.14 compares the deviation from the analytic integral for the sine function in double logarithmic scale for the trapezoidal and parabolic methods as functions of the resolution (number) of sub-intervals. The points represent the numerical integration results over a constant integration region, the lines represent the functions an^{-2} and bn^{-4}, with a and b chosen in such a way that both lines coincide with the numerical error for the smallest number of intervals. The further behavior of both functions and the points confirms the expected dependence on n.

This example should demonstrate to you how versatile already the Taylor series of the fifth order is and, therefore, why we have treated it in such depth.

5.10   Series expansion (2): the Fourier series

5.10.1   Taylor series and Fourier series

The partial sums of the Taylor series approximate a function $f(x)$ in the vicinity of the computation point x_0 via partial sums of a power series. If it is necessary to approximate a function over a larger interval, one would need terms of a very high order. The polynomial obtained by truncating the Taylor series would have to have as least as many turning points as the function. For periodical functions this would be very tedious for intervals larger than the period.

Periodical functions have great practical importance in telecommunications and electrical engineering. For such functions, approximation using the superposition of *periodical* standard functions (sine and cosine) is much better suited. One expands the function into a series that consists of the fundamental tone and the overtones, i.e. of the functions $\sin nx$ and $\cos nx$ with integer values of n.

The analogy to the analysis of a vibrating string is immediately obvious: $\sin x$ describes the vibration of the fundamental tone, $\sin 2x$ that of the octave, $\sin 3x$ that of the fifth above the octave and so on. For a string that is fixed at both ends x is twice the string length. The variable x is now the product ωt of the angular frequency ω and the time t.

$$x = \omega t = 2\pi\nu t = 2\pi\frac{t}{T}; \quad \nu \text{ frequency of oscillation;}$$

$$T \text{ duration of one period.}$$

Depending on the shape of $f(t)$, one superimposes more or fewer of these sine/cosine oscillations of a certain amplitude. The set of amplitudes of the overtones, i.e. the coefficients of the series expansion, represents the *spectrum* of the periodical oscillation. Spectrum and oscillation forms are corresponding representations of the same phenomenon. This representation in terms of superimposed sine and cosine functions is called the *Fourier series* of $f(t)$.

While the partial sums of the Taylor series approximate the function *in the proximity of a point*, the partial sums of the Fourier series are *approximations for the entire interval of the fundamental period* and therefore also – because of the periodicity of the functions considered – for an unlimited region of the variable x. The Fourier series does not have to coincide with the function at a specific computation point, unlike the Taylor Series, which must coincide with the function at the computation point.

It depends on the properties of $f(t)$ how many overtones need to be superimposed to approximate the function at nearly all points. If one does not interpret the notion of convergence strictly, then Fourier series converge for all functions, even for noncontinuous ones. The convergence is then not necessarily monotone, i.e. it can be better for some values of t and worse for other values of t, and may even fail for some values! At discontinuities one observes overshooting even for higher orders of the series. This is called *ringing* in telecommunications.

Since the periodical phenomena that we consider here are mostly oscillations in time, the variable is usually $x = \omega t$. To also model the phases of the individual overtones, we use a sum of terms with $\sin nx$ and $\cos nx$. The sum then represents a phase-shifted sine or cosine function. Thus the general Fourier series reads:

$$f(t) = \frac{a_0}{2} + \sum_{n=1}^{\infty} a_n \cos(n\omega t) + b_n \sin(n\omega t).$$

For a given spectrum $a_0, a_i, b_i, i = 1, 2, \ldots$, one can calculate $f(t)$. For a given function $f(t)$, all coefficients can be determined and thus the spectrum is known.

5.10.2 Determination of the Fourier coefficients

How do we now obtain the coefficients a_n and b_n?

For the Taylor series, we made use of the fact that, following differentiation, all terms that still contain the distance x to the point of computation become zero, such that the coefficient of the corresponding *constant* term gives up to a factor the corresponding derivative at the point of computation.

For the Fourier series we instead begin by integrating the product of the function and the overtones $\cos(m\omega t)$ or $\sin(m\omega t)$; $m = 1, 2, 3, \ldots$ over one period T of the fundamental frequency ($m = 1$)

$$\int_0^T \cos(m\omega t) f(t) dt = \int_0^T \cos(m\omega t) \left(\frac{a_0}{2} + \sum_{n=1}^{\infty} a_n \cos(n\omega t) + b_n \sin(n\omega t) \right) dt$$

$$\int_0^T \sin(m\omega t) f(t) dt = \int_0^T \sin(m\omega t) \left(\frac{a_0}{2} + \sum_{n=1}^{\infty} a_n \cos(n\omega t) + b_n \sin(n\omega t) \right) dt.$$

This initially looks a bit complicated; however, it turns out that the integral over the constant, i.e. the first term before the sum symbol, nearly always vanishes, since the integral over a period of cosine or sine is zero. Only for $m = 0$ does one obtain a contribution, since we have: $\cos 0 = 1 = \text{const}$. Therefore the following applies:

$$\frac{a_0}{2} = \frac{1}{T} \int_0^T f(t) dt.$$

In addition, the integral over the product of an overtone m and a second overtone n is zero, if m and n are not equal. This also applies when a cosine and sine function are multiplied, because of the sine functions are odd while the cosine function are even with respect to $x = 0$. Therefore, we are left only with the integrals $\cos^2 nx$ or $\sin^2 nx$, which are both $T/2$. Thus the coefficients can be easily written down, but this requires the determination of integrals, which often necessitates numerical calculations.

$$a_n = \frac{2}{T} \int \cos(n\omega t) f(t) dt; \quad b_n = \frac{2}{T} \int \sin(n\omega t) f(t) dt.$$

The simulation in Figure 5.15 visualizes these circumstances that simplify the calculation of the Fourier coefficients. From a selection field, a product of periodic functions of the general form that we are interested in is chosen: $\cos(mx)(a \cos(nx) + b \sin(nx))$.

With slides, the parameters a and b and the integers m and n can be chosen. The function is drawn in red. After activating the field entitled *integral* the blue integral function is calculated over a period of the fundamental oscillation from 0 to 2π. The final value is the definite integral of interest to us.

As a first step, we convince ourselves that integrals over sine and cosine vanish and that the addition of sine and cosine functions results in a phase-shifted sine or cosine function, whose integral also vanishes. The calculation of the integral for the product of the function defined above, with an overtone of initially unknown order, shows that indeed all contributions vanish except for the one where the overtones are identical and the function type (sine or cosine) is the same. One realizes that the symmetry of the different functions with respect to the midpoint of the period on the x-axis is the

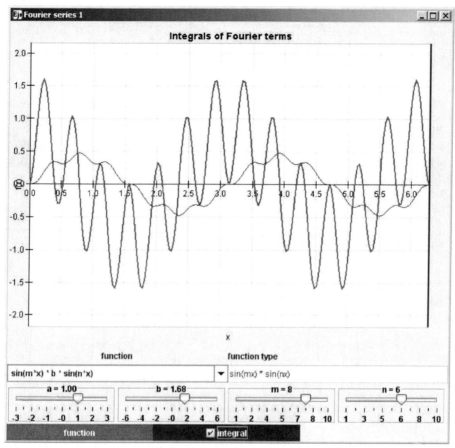

Figure 5.15. Simulation. The simulation visualizes the *orthogonality* of the trigonometric functions. The red curve represents the product of $\cos mx$ and the adjustable overtone $a \cos nx + b \sin nx$; in the figure we have $m = 10$ and $n = 8$. The blue curve shows the integral, whose final value (definite integral over one period of $f(t)$) vanishes for $m \neq n$. For $m = n$ we obtain, when integrating over $a \cos mx \cos mx$, the result $a\pi$, while the integral over the mixed term $b \cos mx \sin mx$ vanishes. The integration is started by selecting the corresponding option box.

reason for this specific result. Thus we have:

$$\int_0^T \cos(m\omega t)\,dt = 0; \qquad \int_0^T \cos(m\omega t)\sin(n\omega t)\,dt = 0;$$

$$\int_0^T \cos(m\omega t)\cos(n\omega t)\,dt = \begin{cases} 0 & \text{for } m \neq n \\ T/2 & \text{for } m = n. \end{cases}$$

This property of the functions sine and cosine means that they are an example of an *orthogonal system of functions*. Two functions are called orthogonal if the following applies:

$$\int_0^T f_1(t)f_2(t)dt = 0 \quad \text{for } f_1(t) \neq f_2(t).$$

In the description pages of the simulation, more detailed instructions and hints for experiments are provided. After opening the simulation you choose the function type and press the enter key. The integration process is animated in order for you to see the difference between the integrals more easily when changing the functions.

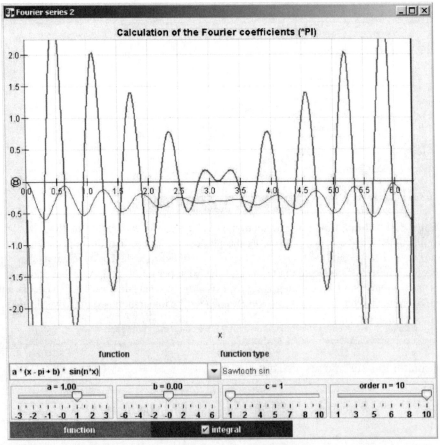

Figure 5.16. Simulation. Calculation of Fourier coefficients for a choice of functions $f(t)$, in the figure for a sawtooth oscillation. $f(t)$. $\sin(10t)$ is drawn in red, and the integral function is drawn in blue; its final value corresponds to the coefficient b_{10} of the 10th *sine overtone*.

5.10.3   Visualizing the calculation of coefficients and spectrum

The simulation in Figure 5.16 visualizes the calculation of the Fourier coefficients for the fundamental tone and the first nine overtones for the following typical periodical functions: *sawtooth, square wave, square-impulse* and *Gaussian impulse*. To this end, the product of the functions under the integral sign is determined and drawn in red while the definite integral is shown in blue. The final value of the integral is, except for a factor π that was suppressed to get more easily readable values, equal to the coefficient of the selected order. The functions are provided with up to three parameters, a, b and c, that control the amplitude, the point of symmetry and the impulse width. From the simulation, the spectra of the functions shown can be obtained in a numerical and experimental manner.

The interactive figure of the simulation shows the situation for the sine coefficients of 10th order of a symmetrical sawtooth. The simulation is started by choosing a function and clicking on the enter key. The description pages and the instructions for experiments contain further details.

5.10.4   Examples of Fourier expansions

In the following interactive examples (Figure 5.17a to Figure 5.17c) the calculation of the coefficients takes place in the background. In the window the function is shown in red and the partial sum of the desired order is shown in blue. The function window is interactive such that many more functions can be entered and a few are suggested in the description. In a text window, the order of the analysis can be adjusted; the approximation order n to be used for the partial sum is selected with a slider. The simulation allows the use of very high orders.

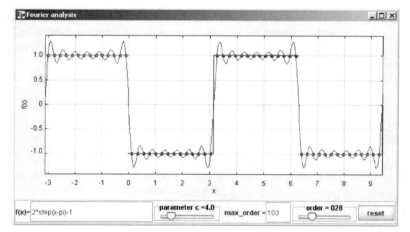

Figure 5.17a. Simulation. Periodical square impulse (red) and its Fourier approximation (blue) of 28nd order. The calculated order n can be chosen.

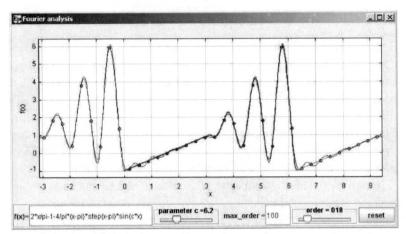

Figure 5.17b. Periodic sawtooth, modulated from the middle of the period by a high frequency sine function (red) and Fourier approximation of 18th order (blue). The modulation frequency can be chosen with the slider.

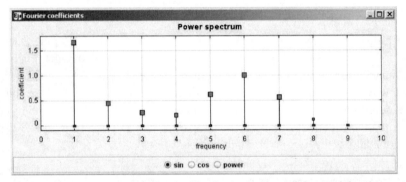

Figure 5.17c. Frequency spectrum for the Fourier expansion of the modulated sawtooth in Figure 5.17b. The abscissa shows the order n of the overtone (fundamental tone $n = 1$); on the ordinate one can choose between displaying the individual coefficients or the total power in a given order.

The calculation of the Fourier expansion of nth order follows immediately after entering the function. The diagram extends beyond the integration region of 2π in order to see the periodic continuation in both directions.

In Figure 5.17a the Fourier expansion of order 43 is shown as an approximation for the symmetrical and periodic square impulse. For the square wave, one recognizes very clearly the typical overshooting at discontinuities, which does not vanish even for very high orders.

In Figure 5.17b, using the same simulation, the approximation of 17th order is shown for a sawtooth oscillation, which has been modulated in a nonlinear fashion with a sine function of high frequency.

The spectrum is shown in a second window of the simulation (Figure 5.17c). It can be changed between sine (a_n), cosine (b_n) and power spectrum ($s_n^2 + b_n^2$). This figure shows the spectrum of the modulated sawtooth, which is rich in overtones and has a pronounced *formant* at the sixth and seventh overtone. In acoustics, formants are defined as limited regions of overtones with large amplitude; they significantly determine the tone quality.

The description of the simulation contains further instructions.

5.10.5 Complex Fourier series

In the space of complex numbers, the Fourier series can be formulated in a very elegant way:

$$f(t) = \sum_{n=-\infty}^{\infty} c_n e^{in\omega t}$$

$$c_n = \frac{1}{T} \int_0^T f(t) e^{in\omega t} \, dt.$$

The connection to the real representation is obtained via reordering the sum and combining, starting with $n = 1$ terms with $-n$ and n. Taking into account $\cos(-x) = \cos(x)$; $\sin(-x) = -\sin(x)$ we get:

$$f(t) = \sum_{p=-\infty}^{\infty} c_n e^{in\omega t} = \sum_{p=-\infty}^{\infty} c_n (\cos n\omega t + i \sin n\omega t)$$

$$= c_0 + (c_1 + c_{-1}) \cos n\omega t + i(c_1 - c_{-1}) \sin n\omega t + \cdots$$

$$f(t) = c_0 + \sum_{p=1}^{\infty} (c_n + c_{-n})(\cos n\omega t + i(c_n - c_{-n}) \sin n\omega t).$$

As a connection between real and complex coefficients we obtain:

$$a_0 = 2c_0; \quad a_n = c_n + c_{-n}; \quad b_n = i(c_n - c_{-n}).$$

The complex formulation is particularly used in electrical engineering. It has the advantage that calculations with exponentials are in general easier and more transparent then those with trigonometric functions.

For the *fast* numerical computation of the components of a Fourier series, a special [FFT] algorithm has been developed, which is known as *FFT* (*Fast Fourier Transformation*).

5.10.6 Numerical solution of equations and iterative methods

In mathematics and physics one often needs to determine the values of a variable, Iteratic
for which a function depending on this variable has certain value C. An identical
problem, as far as the computation is concerned, is to find the value of the variable at
which two functions of one variable have the same value. One solves these problems
by looking for *the zeros of a function*.

> We define $y_1 = f(x)$; $y_2 = g(x)$
>
> For which x is $y_1 = C$?   Answer: $f(x) - C = 0$
>
> For which x is $y_1 = y_2$?   Answer: $h(x) \equiv f(x) - g(x) = 0$.

An analytical solution for finding zeros of a function can only be derived for very
simple functions, thus it is an *exception*. Therefore, one needs a numerical method of
solution that preferably works for all functions and all parameter values.

This is achieved with *iterative methods* that present a *reversal of the question*. One
initially takes a value of the variable, which is probably smaller than the estimated first
zero in the interval of interest, and calculates both the absolute value of the function
value and its sign. Then one increases the variable by a given interval (one can of
course also start from the right and decrease the variable step by step). If the sign
changes one has obviously crossed a zero. Now the direction of the movement is
inverted and the step width is multiplied by a factor < 1. Thus one finds boxes of
decreasing size containing the zero until the deviation of the function value from zero
becomes less than a predetermined tolerance. Then one continues with the process in
the original direction, until all zeros have been found or until a certain threshold for
the value of the variable or of the function itself has been exceeded, and thus one is
outside the region of interest.

For this iteration process, ready made algorithms are available in standard numerical computer codes, which include further refinements. Thus one can, for example,
vary the width of the iteration intervals such that the character of the function is taken
into account. For example with the *Newton* method one uses its slope *the first derivative* to adjust these intervals. Given the speed of today's computers, these refinements
are no problem for simple tasks. The following interactive example in Figure 5.18
determines the zeros of a *function that can be entered at will*. This function is preset
as a polynomial of fourth degree with irrational roots.

The sequence shows the progression of a very simple iteration algorithm. The speed
can be adjusted. The starting point of iteration (magenta) can be dragged with the
mouse. The iteration proceeds with a constant step width to larger x-values until the
sign of the function changes. The initial value is reset to the last value before the sign
change and the step width is decreased by a factor of 10 and the progression to larger
x-values is resumed. This is repeated until the deviation of the y-value from zero
falls below a given tolerance. In the simulation one can choose whether it stops after

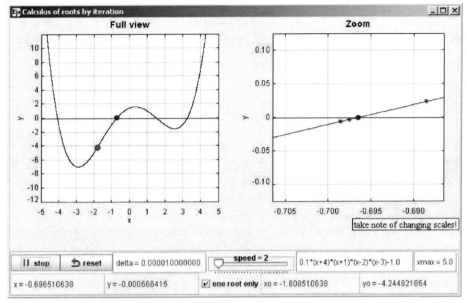

Figure 5.18. Simulation. Animated iterative calculation of the zeros of a function; a polynomial of fourth degree is shown in the figure. The left window shows the whole calculation interval, the right, a section whose scale conforms to the resolution achieved. The last iteration point is shown in blue in both windows while the three predecessors are shown in red in the "looking glass" window. The image shows a return after dividing the interval by 10. The magenta point is the starting point of the iteration. It can be drawn with the mouse. The desired precision *delta*, the number of time steps per second (speed) and the abscissa range x_{max} can be chosen. In the number fields the coordinates of the current iteration point x, y and the initial point x_0, y_0 of the iteration are shown. In the formula window, any functions can be entered whose zeros are to be calculated.

reaching a certain accuracy, or whether all zeros in the variable interval are determined in sequence. In a single calculation the magenta point jumps to the calculated value, while the blue dot shows the first iteration value when determining multiple zeros.

To enable you to follow the progressive iteration with a high level of accuracy, a section of the window is shown in detail like in a magnifying glass, and the scale adjusts to the increasing accuracy.

From the zoom window of Figure 5.18 you can see that the curve is always nearly linear close the root of the curve. The *regula falsi* uses, as the next iteration value for x, the intersection of the secant formed from the two previous iteration points with the x-axis. It therefore leads quickly to the solution. We have, however, chosen the constant step width so that the process can be observed more easily.

Further details and hints for experiments can be found on the description pages of the simulation.

6 Visualization of functions in the space of real numbers

In this chapter simulation is used to represent different types of functions graphically and to visualize them in their two or three-dimensional context. In most cases, the functions have up to four parameters that can be varied.

Physical quantities, such as mass or length, are always associated with a dimension. It is our goal to convey an impression of the *character* and of the *type* of the functions $y = f(x)$. If one considers functions of physical quantities A, for example temperature T, voltage U and mass M, one has to make the argument of the function dimensionless, as a rule. Thus the x in $f(x)$ is then understood as T/K, U/V, M/kg and so on, where K stands for Kelvin, V for Volt and kg for kilogram. The physical quantities appearing in the following section are thus assumed to be made dimensionless in this way. Hint: if the unit is changed then x also changes, for example we have $L/cm = 100L/m$.

For some simulation files, in particular those of functions of three variables, one parameter is changed periodically as a function of time. The animation achieved in this way enhances the spatial impression and rapidly conveys a sense of the influence of that parameter. Animation is also used for the representation of parameter functions as paths on the plane and in space.

Each file contains a description and hints for experiments.

In a selection window, a large number of standard functions $y = f(x)$ are listed according to their *type* (for example *Poisson distribution*, *surface wave*). In a text window the *formula* of the selected function is shown and can be edited or even rewritten from scratch. Changes to the formula can be confirmed with *enter*.

The command panel below the plot mainly looks the same for all simulations; with selection window and formula display, four sliders for adjusting the parameters either continuously or as integer values, input fields for scales etc. and option fields where needed for showing or suppressing additional functions such as the derivative or integral. In the following first example this is discussed in detail.

6.1 Standard functions $y = f(x)$

The following simulation in Figure 6.1 is a *plotter* for arbitrary functions $y = f(x)$. From a selection menu you can choose preset functions, which can be changed. You can also enter totally new analytic functions.

The original function itself is shown in red. With option switches you can call several functions to calculate and display them: *inverse function, first derivative, second derivative* and *integral*.

The derivatives are calculated in secant approximation, the integrals in parabolic approximation.

Inverse function: $x = g(y)$. This involves the problem of finding, for a given function $y = f(x)$ for a certain image variable y, the preimage x. Graphically this corresponds to a reflection of $y = f(x)$ on the angle bisector $y = x$ or swapping of x and y. This line is shown in the corresponding plot (note that in this plot x and y scales are not identical). In the following figure this is shown for a polynomial of fifth order with three zeros. The angle bisector is shown in gray, the inverse function in light brown. The plot of this function is an example of the situation in which the function $y = f(x)$ is unique, each x_i is mapped to exactly one y_i, but the inverse function is not unique, i.e. there are many y_i for which three x_j exist.

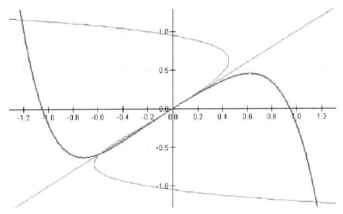

First derivative: $y = \frac{d}{dx} f(x)$ is shown in magenta.

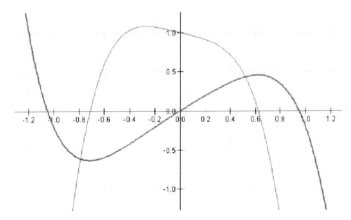

Second derivative: $y'' = \frac{d^2}{dx^2} f(x) = \frac{d}{dx} y'$ is shown in green.

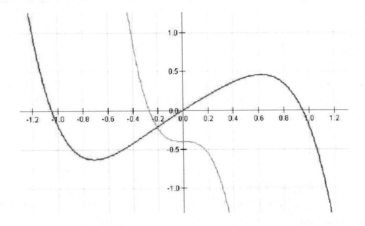

The **integral** $\int_{x_{min}}^{x} f(x)dx$ with adjustable initial value I_0 for x_{min}, is shown in blue.

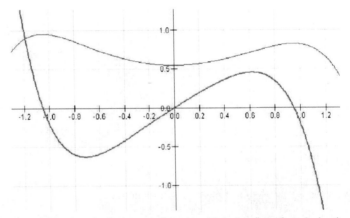

Figure 6.1. Plot of function (red), inverse function (light brown), first derivative (magenta), second derivative (green) and integral (blue), shown for the example of the polynomial of fifth degree $y = -x^5 - 0.2x^2 + x$.

When calculating the integral it is important to remember that the calculation starts at x_0 with an initial value I_0. The variable region and the initial value must be chosen in such a way that the integral curve stays in the window.

Figure 6.2 shows the Gaussian $y = e^{-x^2}$ with inverse function, first and second derivative and integral.

The command panel allows for up to three parameters a, b and c, a continuous variation, and for a fourth parameter, a choice of integers.

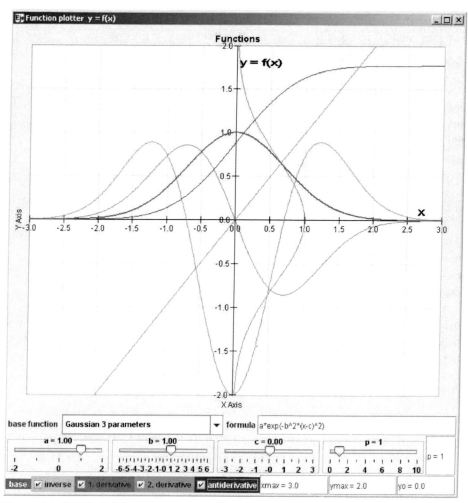

Figure 6.2. Simulation. Function plotter for functions that can be specified at will; optionally, the inverse function (light brown), the first derivative (magenta) and second derivative (green) and the integral (blue) are drawn as well. The figure shows the example of a Gaussian, whose amplitude, width and center can be adjusted with the sliders. The function can be edited.

With the colored option boxes, the inverse function, 1st and 2nd derivative and integral, can be shown or suppressed.

The presentation shows the abscissa and ordinate symmetric to the origin. In the first white field on the bottom you can adjust the variable range $-x_{max} < x < x_{max}$ by hand, in the second field the y-range and in the the third, the initial value of the integral for $-x_{max}$. If the symmetric range of variation is not sufficient for your function, you can increase or decrease it by entering factors in the formulas.

As usual, the window can be pulled to full screen size and, after marking a point, you may read off its coordinates on the lower boundary of the graphic.

The preset functions are:

Functions	Formulas in Java-syntax
constant	a
pth power, $p > 0$ and integer	$a*x\char`^p$
bth power, (b rational; $x > 0$)	$a*x\char`^b$
sine	$sin(x)$
cosine	$cos(x)$
sine with three parameters	$a*sin(b*x + c)$
cosine with three parameters	$a*cos(b*x + c)$
power of sine	$sin(a*x)\char`^p$
power of cosine	$cos(a*x)\char`^p$
tangent with three parameters	$a*tan(b*x + c)$
exponential function	$a*exp(x/b)$
exponential decay	$a*exp(-x/b)$
natural logarithm	$ln(x/a)$
hyperbolic sine	$(exp(a*x) - exp(-a*x))/2$
hyperbolic cosine	$(exp(a*x) + exp(-a*x))/2$
hyperbolic tangent	$(exp(a*x) - exp(-a*x))/(exp(a*x) + exp(-a*x))$
Gauss distribution with three parameters	$a*exp(-b*(x - c)\char`^2$
$(sin x)/x$	$sin(a*x)/(a*x)$
$((sin x)/x)^2$	$(sin(a*x)/(a*x))\char`^2$

In the simulation you may change the preset functions or enter new formulas from scratch.

6.2 Some functions $y = f(x)$ that are important in physics

The following simulation shown in Figure 6.3 uses the basic structure of the previous example.

In this simulation, some important formulas of physics of the type $y = f(x)$ are shown, whose parameters have been chosen in such a way that the variable x and the adjustable parameters correspond to simple, physical quantities. In the second column of the following table, the well-known formulas from physics are given and the formulation in the simulation syntax is given in the second line. Calling the function *random(n)* creates a random number between 0 and n. A random distribution with maximum deviation that is symmetric to zero is obtained as *random(n)* $- n/2$.

In the third column, the meaning of the corresponding variable x and the parameters used are given.

Gaussian, area normalized to 1	$\frac{1}{\sigma\sqrt{\pi}}e^{-(\frac{x-x_0}{\sigma})^2}$ $1/(a*sqrt(pi))$ $*exp(-((x-b)/a)^2)$	$a = \sigma$: standard deviation b: symmetry variable	Gaussian
Gaussian with additive noise	$\frac{1}{\sigma\sqrt{\pi}}e^{-(\frac{x-x_0}{\sigma})^2} + \text{noise}$ $1/(a*sqrt(pi))$ $*exp(-((x-b)/a)^2)$ $+ random(c/10) - c/20$	$a = \sigma$: standard deviation b: symmetry variable $c/10$: maximum added noise	
Gaussian with multiplicative noise	$\frac{1}{\sigma\sqrt{\pi}}e^{-(\frac{x-x_0}{\sigma})^2}(1 + \text{noise})$ $1/(a*sqrt(pi))$ $*exp(-((x-b)/a)^2)$ $*(1 + random(c/10) - c/20)$	$a = \sigma$: standard deviation b: symmetry variable $c/10$: maximum multiplicative noise	
Poisson distribution	$\frac{(x+x_0)^p e^{-(x+x_0)}}{p!}$ $(x + 10)^p$ $*exp(-x - 10)/faculty(p)$	$x + x_0$: expectation value of p $p = 1, 2, 3, \ldots$	Poisson
amplitude modulation	$\sin(\omega_1 t)\cos(\omega_2 t)$ $a*sin(10*x)*cos(b*x)$	$x = \omega t$: angular frequency $10x$: carrier frequency bx: modulating frequency	
phase modulation	$\sin(\omega_1 t + \cos(\omega_2 t))$ $a*sin(5*x + cos(2*b*x))$	$x = \omega t$: angular frequency $5x$: carrier frequency $2bx$: modulating frequency	
frequency modulation	$\sin(\omega_1 t \cdot \cos(\omega_2 t))$ $a*sin(5*x*cos(b/10*x))$	$x = \omega t$: angular frequency $5x$: carrier frequency $b/10x$: modulating frequency	
special theory of relativity: length change	$\sqrt{1 - (\frac{v}{c})^2}$ $sqrt(1 - x^2)$	$x = \beta = v/c$ v: velocity c: speed of light	Relativ
special theory of relativity: mass change	$\frac{1}{\sqrt{1-(\frac{v}{c})^2}}$ $1/sqrt(1 - x^2)$	$x = \beta = v/c$ v: velocity c: speed of light	
Planck's radiation law	$\frac{2\pi hc^2}{\lambda^5}\frac{1}{e^{hc/\lambda kT}-1}$ $a*23340/(x + 2)^5$ $/(exp(8.958/((x + 2)*b)) - 1)$	$x + 2$: wavelength λ in μm a: scale factor b: temperature in 1000 Kelvin	Planck

For calculating the factorial $p!$ this file contains some special code; in other simulation files this function cannot be used.

Figure 6.3 shows a normalized Gaussian impulse with additive noise superimposed on it, and its integral, which in spite of the perturbation reaches 1 quite smoothly and accurately. The formula field can be edited, such that functions can be changed or other functions can be filled in.

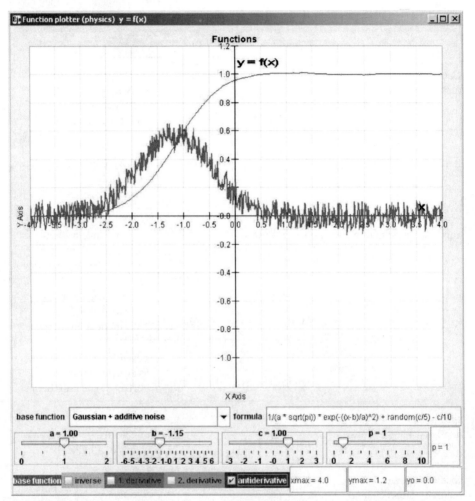

Figure 6.3. Simulation. Function plotter for some physically interesting functions $y = f(x)$. The figure shows a normalized Gaussian (definite integral $= 1$) with superimposed noise and its integral function. Moving the slider p creates a new noise distribution.

6.3   Standard functions of two variables $z = f(x, y)$

For the illustration of surfaces in space $z = f(x, y)$, simulations are particularly useful. Because of the amount of accumulated data, a numerical calculation of the graphs by hand is virtually impossible. In addition, the EJS method makes it possible to rotate the calculated two-dimensional projections of three-dimensional surfaces around any spatial axis by simply dragging the mouse to create a lively three-dimensional impression. If in addition another parameter, for example the extent along the z-axis, is changed periodically (i.e. $z = a \cos(pt) \cdot f(x, y)$), one experiences something close to seeing three-dimensional objects.

The command panel of Figure 6.4a contains four sliders for changing continuously adjustable parameters. The parameter p determines in general the velocity of animation. With the *play* button one starts the animation and it is halted with the *stop* button; the small text field shows the time. *Reset* returns all parameters to their original values.

In the selection field a preset function type can be chosen, whose formula is shown in the formula field below. The term $\cos(t)$ determines the animation in z-direction. You can edit these formulas or enter new ones from scratch (you must not forget to press *enter* to confirm changes!). Figure 6.4a shows a hyperbolic saddle as example.

In the following interactive figures, which show examples of the 3D function plotter, the simulation controls have been suppressed, which correspond to those of Figure 6.4a.

For the plots, the xy-plane $z = 0$ (light brown) was superimposed on the respective spatial surfaces; the origin is in the middle of this surface. The xy-plane can be switched on or off with the option box *show xy plane*. The scales on the axes are all equal and symmetric. You can create different scales via factors in the formulas. The colored points on the z-axis mark the minimum and maximum values of the presentation.

$z = f(x, y)$ only allows for parts of a closed surfaces in space, for example the sphere that is shown here, to be plotted (for example half a sphere). This corresponds to the statement that, in the plane, a function $y = f(x)$ can only represent half a circle. To describe the full circle $y_1 = \sqrt{r^2 - x^2}$; $y_2 = -\sqrt{r^2 - x^2}$ one requires two functions in this representation. If the functions do not yield real values for z, $z = 0$ is shown in the simulations.

You may choose from the functions defined in the table on p. 117. The list of formulas also gives the syntax that must be adhered to when editing.

You can use this file to train your spatial sense and to study the meaning of specific equations, while at the same time having ample leeway to come up with your own formulas. You may also study the influence of the signs and the powers appearing in the formulas. If the uniform scaling used proves inconvenient, for example when dividing by 0, you may adjust the scaling in the formulas accordingly with additive or multiplicative constants in the formulas.

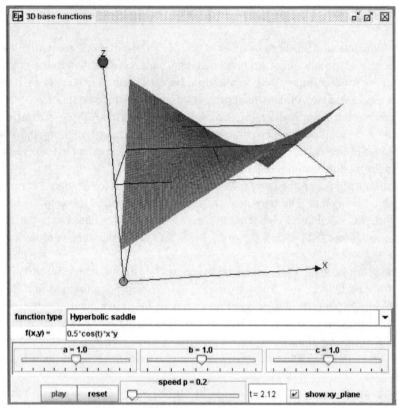

Figure 6.4a. Simulation. Function plotter for animated spatial surfaces $z = f(x, y)$; in the figure a hyperbolic saddle is shown. Up to three parameters, a, b, and c, can be adjusted with sliders. The animation velocity is adjusted with the slider p. The xy-plane can be shown or suppressed.

Further instructions can be found in the description pages of the simulation.

The EJS 3D projection offers, in the *active simulation*, many possibilities for visual representation. We show this in the following non-interactive static pictures for the example of the elliptic–hyperbolic saddle.

Default picture: when calling up the simulation you see, as in Figure 6.4a, the projection of the spatial surface with a xyz-trihedron in a preset *perspective*, with the more distant lines pictured smaller than those that are closer.

Rotation: With the mouse one can dock on to any of the axes and rotate the projection at will.

Shifting: When pressing the *ctrl* key, you can move the representation of the projection surface with the mouse and position it as desired.

Zoom: When pressing the *shift* key you may blow up or shrink the representation by pulling with the mouse. You also may switch or pull the root window to full screen size.

Functions	Formula in Java-syntax of the simulation
plane in space	$cos(p*t)*((b*x) + (a*y)) - c$
paraboloid of revolution	$a*cos(p*t)*(x\hat{\ }2 + y\hat{\ }2) - c$
general paraboloid	$cos(p*t)*((b*x)\hat{\ }2 + (a*y)\hat{\ }2) - c$
parabolic saddle	$cos(p*t)*((b*x)\hat{\ }2 - (a*y)\hat{\ }2) - c$
sphere	$sqrt((a)\hat{\ }2*abs(cos(p*t)) - x\hat{\ }2 - y\hat{\ }2)$
ellipsoid of revolution	$sqrt((b*c)\hat{\ }2*abs(cos(p*t)) - ((c + 1)*x)\hat{\ }2 - (c*y)$
general ellipsoid	$sqrt(a*b - b*x\hat{\ }2 - a*y\hat{\ }2)$
hyperboloid of revolution	$sqrt(a*cos(p*t)\hat{\ }2 + x\hat{\ }2 + y\hat{\ }2) - c$
general hyperboloid	$sqrt(a\hat{\ }2 + b*x\hat{\ }2 + c*y\hat{\ }2) - p$
elliptic hyperbolic saddle	$sqrt(a\hat{\ }2 - cos(p*t)*(b*x\hat{\ }2 - c*y\hat{\ }2))$
hyperbolic saddle	$cos(p*t)*x*y$
standing wave	$a*(sin(pi*x + p*t) + sin(-pi*x + p*t))$
radial surface wave	$a*sin(pi*(x\hat{\ }2 + y\hat{\ }2) - p*t)/sqrt(0.1 + x\hat{\ }2 + y\hat{\ }2)$
(decay like $1/r$)	

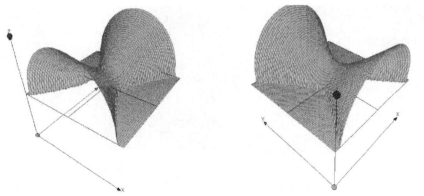

Figure 6.4b. Choice of different viewing directions for perspective distortion, shown for an elliptic–hyperbolic saddle of the simulation in Figure 6.4a. The object can be turned by pulling the arrow inside the spatial trihedron with the mouse. With the perspective representation, the more distant lines are pictured smaller than equally long lines that are closer.

Further special perspectives: are obtained with a *context menu* that appears when pressing the right mouse button on the plot. In the upper line you follow the entries **elements option/drawingPanel3D/Camera** and the following **Camera Inspector** appears (see Figure 6.4c).

You may chose the following options with the projector:

No perspective: The presentation now does not show perspective distortion for the same projection (Figure 6.4d).

On the xy-, yz- or xz-plane: Here you see the projection align an axis, namely the one that is not mentioned (Figure 6.4e).

Camera Inspector	_ □ ×
Projection mode	

⦿ **Perspective**   ○ Planar XY   ○ Planar YZ

○ No perspective   ○ Planar XZ

Camera parameters

X	-0.768	**Focus X**	-0.350
Y	-9.796	**Focus Y**	-0.140
Z	2.458	**Focus Z**	-0.500
Azimuth	-1.614	**Rotation**	0.000
Altitude	0.297	**Screen**	12.500

Suggested camera settings

Figure 6.4c. Camera inspector, which is called with the right mouse key from the context menu. One can choose between different perspectives and projections, enter the parameters of a special projection as numbers, and return to the original state.

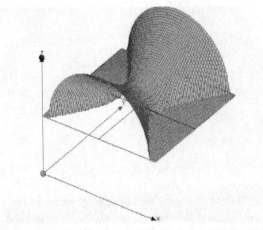

Figure 6.4d. Presentation without perspective distortion.

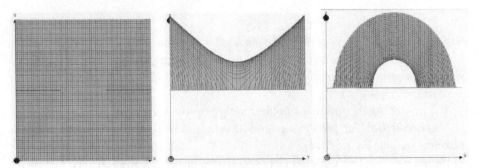

Figure 6.4e. Projections along the three axes.

For the different representations, the optimal visualization depends on the parameters used, which need to be changed when adjusting the representation.

Reset Camera resets the *Camera inspector* to a simple perspective. This is useful if you have created a perspective, that is too confusing. Alternatively, you may switch to another function and then back and recalculate the plot with the original parameters.

6.4 Waves in space

With the function plotter described above, waves in space can be presented quite vividly. One or more space variables then appears in a periodic function, for example as $\cos x$. The spatial surface will then be periodic in one or two dimensions. In the simulation for Figure 6.5, a number of such waves are preset.

We daily observe surface waves in a multitude of shapes on water. In general, these waves propagate in time in one direction without changing their character noticeably in small regions of space. In the simulation, this can be reproduced by adding a phase pt and incrementing the time t continuously and evenly: $\cos(x - pt)$. The wave that is stationary for $p = 0$ moves for $p > 0$ with constant velocity in the positive x-direction. The propagation velocity is set with p. This animation makes the projection picture of the wave very vivid.

The following functions are preset in the selection field:

Functions	Formula in simulation syntax
plane wave in x	$a*sin(b*x - p*t)$
plane wave in y	$a*sin(b*y - p*t)$
plane wave with arbitrary direction	$0.3*sin(6*pi*a*(b*y + c*x)/sqrt(b*b + c*c) - p*t)$
concurrent interference f_1	$a*(sin(b*y - p*t) + sin(b*y - p*t))$
opposing interference f_1	$a*(sin(b*y - p*t) + sin(-b*y - p*t))$
concurrent interference $f_1 + f_2$	$a*(sin(b*y - p*t) + sin(c*y - p*t))$
opposing interference $f_1 + f_2$	$a*(sin(b*y - p*t) + sin(-c*y - p*t))$
orthogonal interference $f_1 + f_2$	$a*(sin(b*x - p*t) + sin(c*y - p*t))$
concurrent interference, adjustable angle c	$a*(sin(b*(y - (c - pi)*x) - p*t)$ $+ sin(b*(y + (c - pi)*x) - p*t))$
opposing interference, adjustable angle c	$a*(sin(b*(y - (c - pi)*x) - p*t)$ $+ sin(b*(-y + (c - pi)*x) - p*t))$
diverging radial wave	$a*sin(b*(x*x + y*y) - p*t)$
converging radial wave	$a*sin(b*(x*x + y*y) + p*t)$
stationary radial wave	$a*(sin(b*(x^2 + y^2) - p*t)$ $+ sin(b*(x^2 + y^2) + p*t))$
diverging surface wave	$0.4*a*sin(b*(x^2 + y^2) - p*t)/sqrt(0.1 + x^2 + y^2)$
diverging space wave	$0.2*a*sin(b*(x^2 + y^2) - p*t)/(0.1 + x^2 + y^2)$

The interference of waves with the same direction of propagation is referred to as *concurrent interference*, that of opposite direction as *opposing interference*. We also give examples for the interference of waves of the same frequency as well as of waves of different frequencies, and finally the interference of waves under 90 degree and under adjustable angles.

For radial waves, the simple radial wave with constant amplitude is physically not possible; it is a unrealistic fiction. This is because the amplitude will decay as a function of the radius (distance from the excitation center), since the excitation energy is distributed over a wider and wider circle. For the spatial radial wave, for example the spatial compression wave originating from a nearly point-like source, the section of the excitation is shown in the xy-plane; here the amplitude decays with the radius, i.e. like $1/r^2$, since the energy is distributed over a spherical surface.

With this simulation you can train your spatial awareness for wave phenomena and the corresponding understanding of formulas. When editing the formulas, you can explore many possible ways of simulating natural phenomena. Remember that you may also choose the velocity of propagation differently when superimposing several waves and thus observe the phenomenon of *dispersion*. Further instructions can be found in the description pages.

These animations start in a state of motion. You also may change parameters while the animation is running and switch between function types. Figure 6.5 shows as an example a radial wave in space.

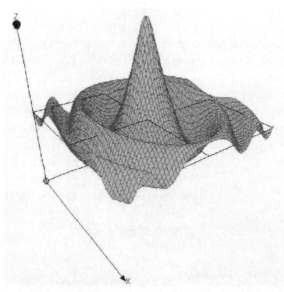

Figure 6.5. Simulation. Function plotter for propagating waves in space. The figure shows a diverging spatial wave excited at the origin.

6.5 Parameter representation of surfaces: $x = f_x(p, q); \; y = f_y(p, q); \; z = f_z(p, q)$

Using the parameter representation it is possible to describe very complicated surfaces in space. The functions f_x, f_y, f_z displayed in the three function windows of the simulation map the pq-plane into the space described by x, y, z. If there are periodic functions of the parameters among f_x, f_y, f_z, closed or self penetrating surfaces in space are created.

From the formula for the first surface in the list of functions, you realize that the parameter v periodically modulates the value z_i of the z-function: $z = z_i a \cos(vt)$. For $t = 0$ the modulation factor is equal to 1. The parameter a determines the amplitude of the modulation; $a - 0.6$ fixes a reasonable initial value. The remaining parameters b and c are not used in this example; please observe for the individual functions which quantities are modulated by a term containing $\cos vt$.

The scale for the x, y and z-axes is adjusted in such a way that the interval $-1 \le x, y, z \le +1$ is covered. The range of the parameters p and q is from $-\pi$ to $+\pi$, such that the simple trigonometric functions like $\cos p$ run through a full period in the parameter interval.

By clicking on the selection window, the preset functions are called.

With the sliders $a, b,$ and c you can also change the parameters of the spatial surfaces during the animation. By editing the corresponding formulas you can also switch the animation to other quantities.

You can edit the formulas in the formula window or enter formulas from scratch. Do not forget to press the *enter* key after doing this.

Some elementary surfaces have already been covered by the basic functions $z = f(x, y)$; thus you may compare the formulas in both representations.

Since p and q are scaled by pi (π), there always appears a factor of $1/pi$, when p and q are directly connected with x, y, z, i.e. outside periodic functions. A factor $\cos vt$ shows that the quantity that is multiplied by it is modulated in the animation. *Reset* returns the value of $\cos vt$ to 1.

The following functions are preset in the selection windows (for the sake of clarity we have left out the multiplication sign * in the simulation syntax).

Tilting plane $x = p/pi; \; y = q/p; \; z = \cos(vt)(a/pi - 0.6)p$

Hyperbolic saddle $x = p/pi; \; y = q/pi; \; z = \cos(vt)pq/pi\hat{\ }2$

Cylinder $x = \cos(vt)a\cos(p); \; y = b\sin(p); \; z = cq/(2pi)$

Möbius strip $x = a\cos(p)(1 + q/(2pi)\cos(p/2));$

$\qquad\qquad\qquad\qquad y = 2b\sin(p)(1 + q/(2pi)\cos(p/2));$

$\qquad\qquad\qquad\qquad z = cq/(pi)\sin(p/2t)$

Sphere $x = cos(vt)acos(p)abs(cos(q))$;

$y = cos(vt)asin(p)abs(cos(q))$; $z = cos(vt)asin(q)$

Ellipsoid $x = acos(p)abs(cos(q))$;

$y = cos(vt)bsin(p)abs(cos(q))$;

$z = csin(q)$

Double cone $x = a/pi(1 + qcos(p))$;

$y = cos(vt)b/pi(1 + qsin(p))$;

$z = cq/pi$

Torus $x = (a + cos(vt)bcos(q))sin(p)$;

$y = (c + cos(vt)bcos(q))cos(p)$; $z = bsin(q)$

8-Torus $x = (a + bcos^2(q))sin(p)$;

$y = ((cos(vt)\hat{}2)c + bcos(q))cos^2(p)$; $z = 0.6bsin(q)$

"Mouth" $x = (cos(vt)c + bcos(q))cos^3(p)$;

$y = (a + bcos(q))sin(p)$;

$z = bsin(q)$

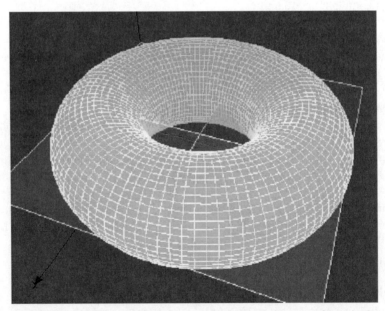

Figure 6.6. Simulation. Function plotter for animated 3D parameter surfaces; in the picture a torus is shown, whose dimensions can be changed with sliders. This animation also contains the *Möbius strip* that was shown at the beginning of the book, in a simpler form.

Boat_1 $x = (c + b\cos(q))\cos^3(p); \quad y = (a + b\cos(q))\sin(p);$

$z = \cos(vt)b\cos(q)$

Boat_2 $x = (c + b\cos(q))\cos^3(p); \quad y = (a + b\cos(q))\sin(p);$

$z = \cos(vt)b\cos^2(q).$

The formulas of the simulation contain additional fixed numbers, which guarantee a reasonable size for the graphs when opening them.

Using the parameter representation, aesthetically very pleasing spatial surfaces can be created, which can be used as an inspiration for design and construction, so that the playful element is not short-changed. The simulation file may now be opened to show the interactive graphic in Figure 6.6 of a *torus*.

The handling of the simulation in Figure 6.6 is analogous to that for the previous 3D presentations. Details and suggestions for experiments are given on the description pages.

6.6   Parameter representation of curves and space paths: $x = f_x(t); \; y = f_y(t); \; z = f_z(t)$

Using this parameter representation, very complicated curves (paths) in space can be described. The functions f_x, f_y, f_z, which are displayed in the three function windows, map the interval covered by the only parameter t uniquely to a curve $x(t)$, $y(t)$, $z(t)$ in space. If f_x, f_y, f_z contain periodic functions of the parameters, closed or self-intersecting space curves are created.

For the simulation in Figure 6.7, the one-dimensional parameter t is interpreted as time. This parameter is repeatedly incremented by a constant time-step, such that the curve starting at the origin grows accordingly, until one of the coordinates becomes larger than 2 and leaves the range of the figure and the animation stops.

The blue path marker is connected to the origin with a vector. The vector and the xy-pane can be switched on and off with the option switch.

The program calculates the functions in time-steps of $\Delta t = p \times 0.1$ milliseconds. Thus animation speed can be set with the slider p. For $p = 0$, the picture is static.

With the sliders a, b and c, up to three constants in the parameter functions can be adjusted between 0 and 1. The sliders actually determine integers $0 < N < 100$, such that the constants $1/N$, as well as the ratio of two of these constants, are rational numbers. This leads to closed orbits in the case of oscillation plots. In the second example the irrational number $\sqrt{2}$ is added to the rational number c, which results in the orbit not being closed. This shows how you can, in general, create orbits that are not closed. You may increase the animation speed to recognize this quickly. For the

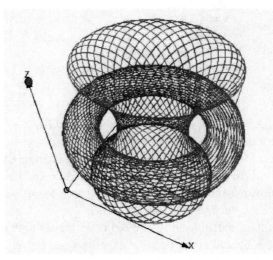

Figure 6.7. Simulation. Function plotter for animated space curves; the figure shows the superposition of a periodic orbit that travels on a hyperboloid and an orbit that travels on a torus.

detailed observation the projection settings of the *camera inspector* are useful. In the xy-plane one sees the corresponding plane orbits, i.e. plane *Lissajous* figures.

Choose after the first animation the constants a, b, c such that the range of coordinates is fully used. Many plots become graphically interesting only if the constants a, b, c are chosen differently. The default value for all of them is 0.5, to show the basic functions during the first run.

You can edit the formulas or enter new ones from scratch.

The scale has been chosen in such a way that, for all three axes, the range -1 to $+1$ is available. The xy-plane is intersected by the z-axis in the middle of the z-vectors. Maximum and minimum values are marked on the z-axis by a red and green point respectively.

With the sliders a, b and c you may, even during the animation, change the parameters of the space curves. With suitable entries of time-dependent functions you can also switch the animation to other quantities.

The handling of the simulation is otherwise again analogous to that of the previous 3D presentations. Details are given in the description pages.

There are, however, two keys for starting the simulation with slightly different functions:

Start   starts the simulation and erases all the curves that are present.
Play   does not delete previous curves, continues for equal parameters with the simulation and superimposes old and new curves for changed parameters or changed function types.

Stop is a second functionality of the *play* button; the simulation can be continued by
pressing *play* again.

Clear deletes all curves.

Reset a b c resets a, b, c to the default values.

This simulation also gives ample opportunities for creative and playful experiments.
Figure 6.7 shows the simulation, with interleaved orbits; in one of these, the hyperbolic envelope is already closed, while in the other, the envelope, in the shape of a
torus, is still open.

7 Visualization of functions in the space of complex numbers

7.1 Conformal mapping

Complex functions $u = F(z)$ map the points z of their domain of definition to points u within their range in the complex plane (to distinguish these functions from real functions we arbitrarily use capital letters for the function):

$$u = F(z).$$

Important complex functions, such as powers, the exponential function and its descendants, among them trigonometric functions and hyperbolic functions, satisfy the property of being *holomorphic*, which means, according to the definition, that they are *complex* differentiable. This means that these functions are differentiable in every point of the complex plane and are also independent of the direction along which one approaches the respective point. Such functions can be differentiated an arbitrary number of times and, therefore, can also be expanded into a power series (Taylor series).

Figure 7.1 from the simulation shown in Figure 7.2a, which will be described shortly, shows how this looks for the concrete case of the mapping $u = z^2$.

The mapping $u = F(z)$ with a holomorphic function is *conformal*, which means *angle-preserving*: curves in the u plane intersect under the same angle as the preimage curves in the z-plane. This is initially baffling, since the shapes consisting of the curves are in general distorted by the mapping.

The left window shows the z-plane, the right one the u-plane. In the z-plane a quadratic grid of points, that lies on parallels to the real and imaginary axis, is shown, which is mapped into the u-plane, undergoing rotation, stretching (for points outside of the indicated unit circle) or compression (for points inside the unit circle) and resulting in a rhombic shape with curved grid lines. In this case, the points on the real axis are transformed to the real axis and therefore the real-valued side of the square remains straight.

On closer examination, it becomes evident that the lines connecting the points in the image plane indeed intersect each other under right angles; the 4 points corresponding to a square of neighboring points in the preimage constitute a square in the image with increasing accuracy for decreasing distance of the points. The conformal angle-preserving property is to be understood in the limit of infinitesimal distances.

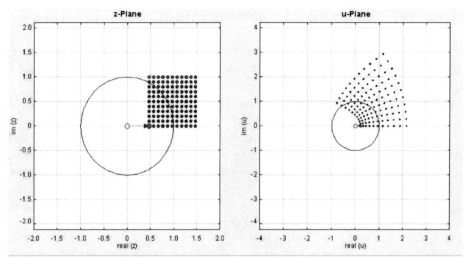

Figure 7.1. Conformal mapping of a quadratic point grid (left window) with the holomorphic function $u = z^2$ to the image plane (right window). Due to stretching, compression and rotation, distortions take place.

The angle-preserving property of conformal mappings is used for practical purposes in engineering, for example to map the solutions of hydrodynamic problems for simple situations to more complex situations. Complex functions are thus not only an abstract mathematical concept, but they have very useful applications.

7.2 Visualization of the complex power function

The following visualization example in Figure 7.2a shows powers for arbitrary positive or negative, integer or fractional, exponents:

Powers

$$u = z^n.$$

Thus we have, for example,

$$u = z^2; \quad u = z^{-3} = \frac{1}{z^3}; \quad u = z^{1.5} = z^{\frac{3}{2}} = \sqrt[2]{z^3}.$$

The control elements for the different simulations of conformal mappings are mostly identical. We describe them in detail for this first example, and refer only to differences later. Extensive details are given in the respective description pages of the simulation.

A quadratic point-grid with preset side length is located in the z-plane. One of the corners is marked in red and connected to the origin by a vector. Using the mouse, the square can be moved in the z-plane while maintaining its orientation by grabbing it on

the red corner. While the other coordinate remain exactly the same, you can change
one of the coordinates with the sliders x, y. Very accurate values can be defined in the
number fields x, y. One can also enter values that go beyond the range of the sliders.
Points with different imaginary parts are differentiated by color in order to be able to
follow the mapping point by point. The color coding can be seen most clearly if you
pull the window to full screen size. The *side length* of the square can be grown or
shrunk to a point using the slider.

In addition a circular color coded point grid is located around the origin with a
default radius that depends on the function. The center of the circle is marked in blue;
it can be moved with the mouse. The points of the circle that are initially on the real
axis and mirror images of each other, are highlighted. The right point that is marked
by a red disk is connected to the origin by a vector, which can be pulled with the
mouse. Using a second slider, the *radius* of the circle can be grown or pulled together
to a point.

By collapsing the square or circle to a point you can plot the other function more
clearly and study the mapping of a single point.

The scale can be adjusted separately in both windows in the number fields *scale_z*
and *scale_u*.

In the u-plane you see the mapping of the individual points of the square or circle
via the chosen function. Accurate coordinates are shown if you click on the points.
An animation is started by pressing the play button, which moves the corner point of
the square arrays step by step.

Even during the animation, the coordinates of the corner points can be changed
with the mouse, the sliders or by entering numbers, so that the whole plane can be
scanned in strips.

With pause/play the animation is stopped. With the initialization button you can
reset the grid, the circle and scale to its original state.

For the power function of Figure 7.2a you may enter an arbitrary positive or negative power n, and also rational numbers. The changes become effective on pressing
the *Enter* button. We have:

$$u = z^n = (re^{i\varphi})^n = r^n e^{in\varphi} = r^n (\cos n\varphi + i \sin n\varphi).$$

The mapping z^n rotates a point $z = re^{i\varphi}$ from the preimage plane to the image plane
by $n - 1$ times its angle. Its absolute value increases to r^n for $n > 1$ and decreases for
$n < 1$. The unit circle is mapped to itself under rotation.

Due to the angular rotation, it follows that for $n > 1$ the simple u-plane is not
sufficient to accommodate the mapping of all z-values. For $n = 2$ or $n = 3$, the
mapping provides for a two-or threefold coverage. In complex analysis, one refers to
n Riemannian sheets of the u-plane. On these sheets the function $u(z)$ is unique: each
point on the z-plane leads to one point on the u-plane. The same is true for the inverse
function $z(u)$. In the simulation picture the Riemannian sheets are superimposed, as

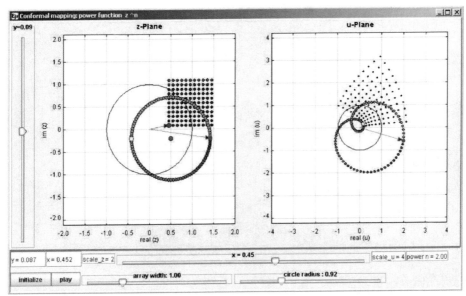

Figure 7.2a. Simulation. Complex power function $u = z^n$; conformal mapping of a point grid and of a circle for $n = 2$. In the left window, for the z-plane, the lower left corner of the grid can be pulled with the mouse and the distance between the points can be adjusted with the lower left slider. The center of the circle can be pulled and its radius can be adjusted with the lower right slider. The yellow point marks the point that has been turned by π in the preimage of the circle. The power n can be chosen at will in the simulation (in the figure $n = 2$).

one can easily see from the mapping of the circle: the two loops belong to different sheets.

For fractional exponents n and negative real values the mapping splits the point grid in two sections, which is initially surprising. On one section lie the transformed points from the positive imaginary half plane, on the second section, the points of the negative imaginary half plane. Whether the u-plane is covered only partially or many times depends on whether n is larger or smaller than 1. Figure 7.2b shows the picture for $n = 0.5$ ($u = z^{0.5} = \sqrt[2]{z}$).

With a bit of calculation, it is easy to see that the splitting has to be as observed for $n = 0.5$, i.e. for $u = z^{\frac{1}{2}} = \sqrt[2]{z}$. The point $z = i$ (angle of $90°$) is mapped to the point $u = \sqrt{i} = \frac{1+i}{\sqrt{2}}$ with the angle of $45°$, as we show by inverting the function:

$$\left[\frac{1}{\sqrt{2}}(1+i)\right]^2 = \frac{1}{2}(1 + 2i + i^2) = \frac{1}{2}(1 + 2i - 1) = i \quad \text{q.e.d.}$$

How is the point $-i$ transformed? We assume that it is mapped to a point that is complimentary to the one obtained above, (same real component, opposite sign of the

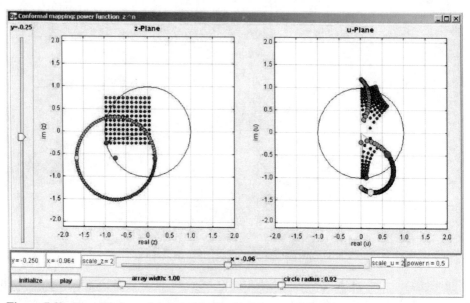

Figure 7.2b. Example from simulation in Figure 7.2a: conformal mapping with the holomorphic function $u = z^{0.5}$ for a point grid on a square and a circle (left window). The z-plane is mapped to the positive real half of the u-plane. The mapping splits the z-plane into two sections for positive and negative imaginary parts.

imaginary component) and again prove this via the inverse function:

$$\left[\frac{1}{\sqrt{2}}(1 - i)\right]^2 = \frac{1}{2}(1 - 2i + i^2) = \frac{1}{2}(1 - 2i - 1) = -i \quad \text{q.e.d.}$$

Thus we indeed have the following situation: the point $-i$ and all other points with negative imaginary components are mapped to the section of the u-plane where the imaginary component is negative and all points with positive imaginary components are mapped to a section with a positive imaginary component, which is its mirror image.

For the circle around the origin, the situation can be most easily understood. For $n = 2$ the circle is mapped to two segments, as soon as individual points have a negative imaginary component. By counting you may convince yourself that there are equally many points on the two partial curves if the setup is symmetric to the origin.

For suitable parameters, the conformal mapping yields very interesting symmetries. Figure 7.2c shows on the left-hand side for the 17th and on the right-hand side for the 60th power, the superimposed mapping of 100 points of a circular array of radius 1 on as many Riemannian sheets. The array is slightly shifted from the origin. Remember that the unshifted unit circle is mapped to itself, and thus the shifted one will be mapped to its immediate vicinity.

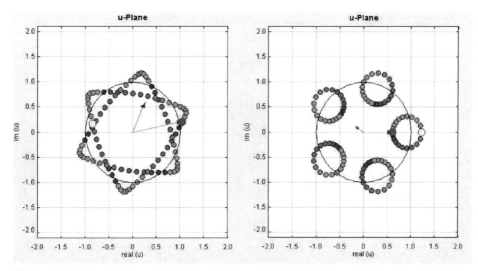

Figure 7.2c. Example from simulation in Figure 7.2a: $u = z^n$ for a circle with radius 1 that has been slightly shifted from the origin. On the left for $n = 17$ and on the right for $n = 60$ with different shift.

The simulation provides many opportunities for experiments, which can be accessed with the interactive simulation in Figure 7.2a. The description pages contain further details and suggestions for experiments.

7.3 Complex exponential function

As second example of conformal mapping, we show the complex exponential function. We generalize it to an arbitrary base a:

$$u = a^z = e^{z \ln a}.$$

With $a = e$ we obtain the normal exponential functions, with $a = \frac{1}{e}$ we obtain the exponential decay function.

$$a = e \rightarrow u = e^x (\cos y + i \sin y)$$

$$a = \frac{1}{e} \rightarrow u = e^{-x} (\cos y - i \sin y);$$

$$\text{(because of } \cos(-y) = \cos(y); \ \sin(-y) = -\sin(y))$$

in general $u = a^z = (e^{\ln a})^z = e^{(\ln a)(x+iy)} = e^{x \ln a} e^{iy \ln a}$

$$= e^{x \ln a} \cdot (\cos(y \ln a) + i \sin(y \ln a)).$$

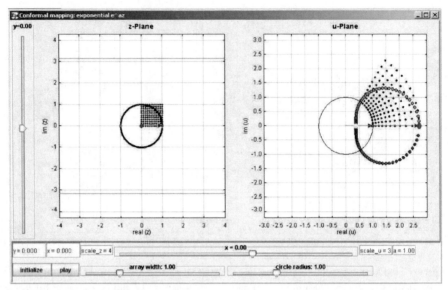

Figure 7.3a. Simulation. Conformal mapping with the complex exponential function $u = e^z$; mapping of a point grid and of a circular array with radius 1 around the origin of the z-plane to the u-plane. The unit circle is drawn in black. *Play* shifts the array along the imaginary axis. The parameter a can be chosen in the number field a.

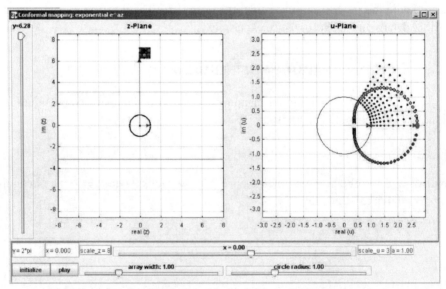

Figure 7.3b. Conformal mapping with the complex exponential function $u = e^z$; Mapping of a circular array with radius 1 and a point grid that has been shifted by 2π along the imaginary axis. The image in the u-plane is identical to the image in Figure 7.3a, where the point grid is located at the origin. The unit circle is drawn in black. The boundaries of a period that is symmetric to the origin are marked in red.

Thus the choice of a base $\neq e$ can be compensated for via a coordinate transformation: $x' = x \ln a; y' = y \ln a$. The simulation uses the same setup as for the power function, and is shown in Figure 7.3a for the simple exponential function with $a = e$.

The real point 1 is mapped to the real point $e = 2.718\ldots$. Negative real parts $x < 0$ of $z = x + iy$ lead to a mapping into the inside of the unit circle, positive ones to a mapping into the outside of the unit circle (marked in the picture by a circle). This is for the following reason:

$$z = e^{i\varphi} \rightarrow u = e^{e^{i\varphi}} = e^{\cos\varphi + i\sin\varphi};$$

however, we have $\cos < 1$ in the range $\pi/2 < \varphi < 3\pi/2$ such that we have $e^{\cos\varphi} < 1$.

The fundamental peculiarity of the complex exponential function is made clear in this simulation: If one moves the point grid along the *imaginary* axis, it is turned in the image plane without additional distortion around the origin and arrives, following a shift by $2\pi i$, at its original position. A strip of the z-plane that is parallel to the real axis of width 2π fills a complete Riemannian sheet in the u-plane. This also shows the periodicity of the trigonometric functions. Figure 7.3b shows the case in which the simulation in Figure 7.3a is shifted by $2\pi i$.

A shift of the grid array along the real axis in positive direction results in exponential expansion, a shift in negative direction in exponential decay.

Interesting results are observed for rational or negative values of a (e.g. $\ln a = -5/3; a = 0.1888\ldots$). The description pages contain additional information.

7.4 Complex trigonometric functions: sine, cosine, tangent

From the complex exponential function, it is only one step to the complex trigonometric function. In addition to the Euler formula $e^{iz} = \cos z + i \sin z$, we require the definitions of the hyperbolic functions sinh and cosh:

$$e^{iz} = \cos z + i \sin z; \ e^{-iz} = \cos(-z) + i \sin(-z) = \cos z - i \sin z;$$

$$\rightarrow \sin z = \frac{e^{iz} - e^{-iz}}{2i}; \ \cos z = \frac{e^{iz} + e^{-iz}}{2};$$

$$\sinh z = \frac{e^z - e^{-z}}{2}; \ \cosh z = \frac{e^z + e^{-z}}{2}; \rightarrow \cosh^2 z - \sinh^2 z = 1;$$

auxiliary results: $\cos z = \cosh(iz); \ \sin z = 1/i \sinh(iz);$

$$\cos(iz) = \cosh(z); \ \sin(iz) = i \sinh z$$

With $e^{iz} = e^{ix-y} = e^{-y}e^{ix} = e^{-y}(\cos x + i \sin x)$

$$e^{-iz} = e^{-ix+y} = e^y e^{-ix} = e^y(\cos x - i \sin x) \quad \text{it follows that}$$

$$\sin z = \sin x \frac{(e^y + e^{-y})}{2} + i \cos x \frac{(e^y - e^{-y})}{2} = \sin x \cosh y + i \cos x \sinh y$$

$$\cos z = \cos x \frac{(e^y + e^{-y})}{2} - i \sin x \frac{(e^y - e^{-y})}{2} = \cos x \cosh y - i \sin x \sinh y$$

$$\tan z = \frac{\sin z}{\cos x} = \frac{\sin x \cosh y + i \cos x \sinh y}{\cos x \cosh y - i \sin x \sinh y}$$

$$= \frac{(\sin x \cosh y + i \cos x \sinh y)(\cos x \cosh y + i \sin x \sinh y)}{(\cos x \cosh y - i \sin x \sinh y)(\cos x \cosh y + i \sin x \sinh y)}$$

$$\tan z = \frac{\sin x \cos x + i \sinh y \cosh y}{\cos^2 x + \sinh^2 y}.$$

7.4.1 Complex sine

When shifting the point arrays parallel to the real axis, one observes their periodic mapping. The square array is then mapped into a region that is bounded by orthogonal ellipses and hyperbolas. Further details and hints for experiments are given in the description pages of the simulation.

7.4.2 Complex cosine

As is to be expected, the mapping via the cosine for a phase shift by $\pi/2$ on the real axis leads to the same result as the mapping for the sine. Figure 7.5 shows this for the same configuration of the u-plane as in Figure 7.4. Further details and hints for experiments are given in the description pages of the simulation.

7.4.3 Complex tangent

In addition to the expected periodicity under shifts parallel to the real axis, the complex tangent shows, because of its divergence with sign change at odd multiples of $\pi/2$, a wealth of interesting phenomena. Because of the high sensitivity close to the divergences you should, in addition to the sliders for the coordinates of the grid array in the z-plane, also use the two number fields, in which exact values for x and y can be entered. They can be chosen outside the intervals covered by the sliders.

Straight lines parallel to the real and imaginary axes are mapped into closed curves around and through the points $+i$ and $-i$. The region with imaginary values larger than π is mapped to the point i, the region with imaginary values smaller than π is mapped to the point $-i$. Further details and hints for experiments are given in the description pages of the simulation.

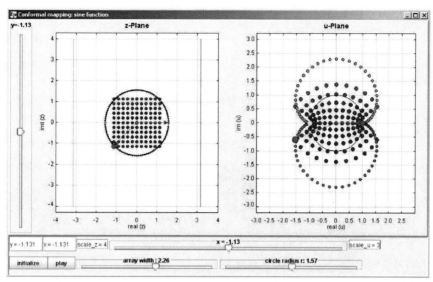

Figure 7.4. Simulation. Conformal mapping with the complex trigonometric function $u = \sin z$; mapping of a point grid and a circular array around the origin with radius $\pi/2$ from the z-plane to the u-plane. The circle with radius $\pi/2$ in the z-plane and the unit circle in the u-plane are drawn in black. In the z-plane, the boundaries of a period are drawn in red. The *Play* button shifts the square array along the real axis.

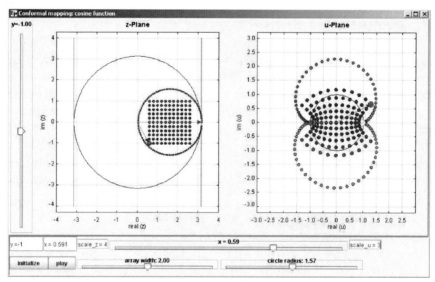

Figure 7.5. Simulation. Conformal mapping with the complex trigonometric function $u = \cos z$; mapping of a point grid that has been shifted by $\pi/2$, relative to the origin, to the u-plane. The circle with radius $\pi/2$ in the z-plane and the unit circle in the u-plane are drawn in black. In the z-plane the boundaries of a period are drawn in red.

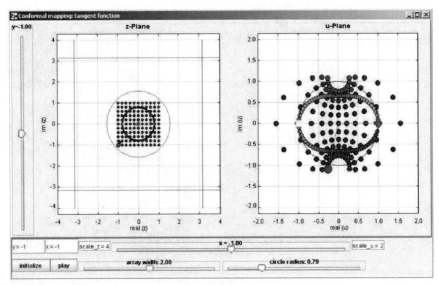

Figure 7.6. Simulation. Conformal mapping with the complex trigonometric function $u =$ tan z; mapping of square point grid and a circular array around the origin of the z-plane to the u-plane. A circle with radius $\pi/2$ and in the z-plane and the unit circle are drawn in black. The boundaries of a period are drawn in red. *Play* shifts the square array parallel to the real axis.

7.5 Complex logarithm

We conclude the chapter on conformal mapping with the natural logarithm. It is well known that there exists no logarithm for negative numbers in the space of real numbers, since the inverse function e^x always leads to a positive number. This limitation is lifted in the space of complex number, in which the logarithm is well defined for all numbers.

To calculate the complex logarithm, one has to use the complex number z in a form that allows for the separation of real and imaginary parts when taking the logarithm. This is not the case for the form $z = x + iy$, but in polar coordinates it works out as:

$$z = re^{i\phi}; \quad r = |z| = \sqrt{x^2 + y^2}; \quad \phi = \arctan \frac{y}{x};$$

$$\ln z = \ln \sqrt{x^2 + y^2} + i(\phi + k2\pi); \quad k \text{ integer}$$

$$\textit{main value for } k = 0: \ \ln z = \ln \sqrt{x^2 + y^2} + i\phi = \frac{1}{2}\ln(x^2 + y^2) + i\phi.$$

Because of the periodicity of the exponential function with a period of $2\pi i$, the z plane is mapped identically to an infinite number of strips parallel to the real axis in the u-plane of width 2π. The main value for $k = 0$ maps the z-plane to $pi < y < \pi$ on the u-plane.

In Figure 7.7 one sees for the quadratic point array the logarithmic compression along the real axis and the compression due to the arc-tangent along the imaginary axis.

For the logarithmic mapping, one distinguishes four regions according to the real component of z in the z-plane:

- $x \geq 1$: for these values the logarithm is positive in the space of real numbers. Complex numbers in this region are transformed to a region with $x > 0$, which is bounded by the green curve in Figure 7.7. The numbers with equal imaginary components lie on curves orthogonal to the green curve and are marked by the yellow line for $y = 1$ in Figure 7.7.
- $x \leq -1$: for these numbers the logarithm does not yield a real solution. Numbers in this region are transformed to regions with $x > 0$ and imaginary parts that lie on the boundaries of the strip. The bounding curves are analogous but shifted and reflected with respect to the first case. An interesting case is $\ln(-1) = 0 + i\pi = i\pi$, the symmetric solutions are $\ln i = \frac{1}{2}i\pi$ and $\ln(-i) = \frac{3}{2}i\pi$.
- $0 < x \leq 1$: here we have, for real numbers, real negative values of the logarithm. Numbers in this region are, depending on the imaginary component, transformed into the positive or negative half of the strip.
- $-1 \leq x < 0$: here the logarithm has no real values. Depending on their imaginary component, numbers in this region are transformed to the negative or positive half plane of the strip and we have $\frac{\pi}{2} < |y| < \pi$ for all y. The bounding curves are continuations of the first case.

A circle around the origin is transformed into a line parallel to the imaginary axis, since the real component of the logarithm $0.5 \ln(x^2 + y^2) = \ln r$ is constant on it. Changing the radius shifts the line in the x-direction.

How are the curves defined, which are shown in Figure 7.7 and appear after activating the switch *visible*? In the z-plane, $x = 1$ is the boundary for positive logarithms. Therefore, the coordinates of the bounding curve in the u-plane are: $x = 0.5 \ln(1 + y_z^2)$; $y = \arctan y_z$. For a line with an imaginary component $y = 1$ in the z-plane, we obtain, in the u-plane, $x = 0.5 \ln(x_z^2 + 1)$; $y = \arctan(\frac{1}{x})$. These two curves are orthogonal to each other.

Further details and hints for experiments can be found in the description pages of the simulation.

This relatively complex example demonstrates quite clearly the advantage of an interactive simulation over a discussion with formulas and words. When moving the arrays parallel or at a right angles to the imaginary axis, the context is immediately grasped visually, which could otherwise only be described in lengthy and time consuming verbal descriptions.

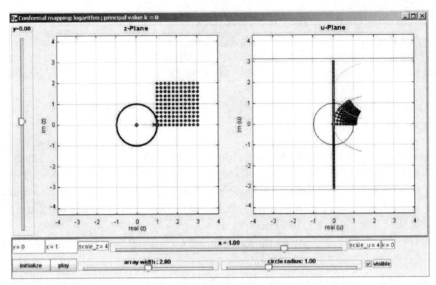

Figure 7.7. Simulation. Conformal mapping with the complex function $u = \ln z$: mapping of a point grid and of a circular array around the origin with radius x in the z-plane to the u-plane. A circle with radius e in the z-plane and a circle with radius π in the u-plane are drawn in black. The red lines in the z-plane mark the boundaries of the main value of the logarithm. *Visible* shows the transformed curves of parallels to the x and y axes in the z-plane. *Play* shifts the array parallel to the real axis.

From the many examples shown above, it should have become clear how to calculate and visualize conformal mappings in general. The examples include, in the custom page of the EJS console, the code for the other functions that fit on a few lines in an inactive mode. In addition the code for cot(z) and for the complex hyperbolic functions $\sinh(z)$, $\cosh(z)$, $\tanh(z)$ and $\coth(z)$ is found there.

From this, it is easy to derive the code for further conformal mappings.

8 Vectors

8.1 Vectors and operators as shorthand for n-tuples of numbers and functions

In secondary school the discussion of functions is mostly restricted to functions of one
variable, i.e. to $y = f(x)$ in Cartesian coordinates or $r = g(\varphi)$ in polar coordinates.
Therefore, one gets used at school to the visualization of functional relationships in
the xy-plane.

Real events cannot be described is this way, since they always take place in three-
dimensional space with coordinates x, y, z or in a four-dimensional continuum, denoted by the space coordinates x, y, z and the time t. As an auxiliary workaround, one
uses only a restricted projection to a plane in space. This is possible if one assumes
that some variables are constant. One example would be $y = f(t)$ for the movement along a straight path that is mapped to the y-axis and, instead of the x-variable,
the parameter t is changing. One can possibly take into account a second quantity x
that is changing in discrete steps, by plotting a family of curves in a plane system of
coordinates, for example $y = f(t, x_i), i = 1, 2, 3, \ldots$.

As soon as one wants to present events in space it becomes more complicated.
The uniform movement of a point mass, i.e. without the influence of any force, requires three "plane" parameter equations, for example $x = at + a_0$; $y = bt + b_0$;
$z = ct + c_0$. If one wants to describe its movement under the influence of a force
that changes from point to point, one requires equations that describe for every point
in space both the absolute values as well as the direction of the force on the moving
body. In coordinate notation this becomes easily messy and not at all vivid.

To come close to the vividness of two-dimensional presentations, one instead uses
a kind of shorthand, which combines the three space components in a *vector* and the
functions connected to it or acting on it in an *operator*. If one combines the three coordinates in the vector X and three functions of time in the operator F, one can combine the above three equations as $X = F(t)$, which is considerably clearer. Whether
it makes sense depends on the specific problem at hand, i.e. on whether the three components of F have a logical connection with each other. This is obviously the case for
the simple movement considered above.

As soon as one starts to substitute numbers and to do calculations with them, one
can find ways of decomposing the relationship into its individual components and
to formulate the corresponding algorithms. However, this process still often greatly
benefits from the symbolic grouping of the individual relationships. Because of the

repeated appearance of the always identical formalisms for physical problems, the formulation often becomes routine.

This approach does not have to be restricted to three-dimensional descriptions, but can, in principle, be extended to an arbitrary number of dimensions. One can for example describe the position of two points in the three-dimensional space via *two* arrows or vectors starting at the origin (x_1, y_1, z_1 and x_2, y_2, z_2) in this space or via *one* vector in the six-dimensional space ($x_1, y_1, z_1, x_2, y_2, z_2$). In quantum mechanics one works with vectors in the infinitely dimensional *Hilbert* space. Plane problems can be described by two-dimensional vectors that can be considered to lie in the complex plane.

Vector algebra and vector analysis, in which partial differentiations take place, are an especially important mathematical tool of theoretical physics and therefore are often treated in depth in many textbooks for first year students. Their objects and operations are not easily accessible to the untrained imagination. Therefore, the following sections concentrate only on the interactive visualization of fundamental aspects.

8.2 3D-visualization of vectors

The classical visual presentation of a vector is an arrow in space, whose length defines an absolute value and whose orientation defines a direction. The place at which the arrow is situated is arbitrary; one can, for example, let it start as a *zero-point vector* from the origin of a Cartesian system of coordinates. Thus its endpoint (the tip of the arrow) is described by the three space coordinates x, y, z in this system of coordinates. Its length a, also referred to as the *absolute value* of the vector, is obtained from the theorem of Pythagoras as $a = \sqrt{x^2 + y^2 + z^2}$.

It obviously does not matter how the system of coordinates, with respect to which the coordinates of the vector are defined, is orientated in space. Under a change of the coordinate system (translation or rotation), the individual coordinates also change, but the position and length of the vector are not affected by this. They are *invariant* under translation and rotation. This property provides the definition of a vector.

Quantities that can be characterized by specifying a *single* number for every point in space are called *scalar*, in contract to vectors; an example would be a density- or temperature distribution.

The three-dimensional zero-point vector represents the position coordinates of a point in space. It is customary to write them as a matrix with only one column or line. As symbols one often uses a_1, a_2, a_3 for the vector **a** or x_{11}, x_{12}, x_{13} for the vector $\mathbf{x}_i$. Thus the following representations are synonymous:

$$\mathbf{a} = \begin{pmatrix} a_1 \\ a_2 \\ a_3 \end{pmatrix} = (a_1, a_2, a_3), \qquad \text{absolute value } |\mathbf{a}| = \sqrt{a_1^2 + a_1^2 + a_2^2}$$

$$\mathbf{x}_1 = \begin{pmatrix} x_{11} \\ x_{12} \\ x_{13} \end{pmatrix} = (x_{11}, x_{12}, x_{13}), \quad \text{absolute value } |\mathbf{x}_1| = \sqrt{x_{11}^2 + x_{12}^2 + x_{13}^2}.$$

Symbols for the vector as a *whole*, such as $\mathbf{a}$ and $\mathbf{x}_1$, were introduced at a time when they were written by hand. Some of the formats used back then, such as cursive letters with an arrow on top, nowadays lead to a somewhat inconvenient typesetting situation, since they cannot be entered quickly on the the PC keyboard. Thus we use,

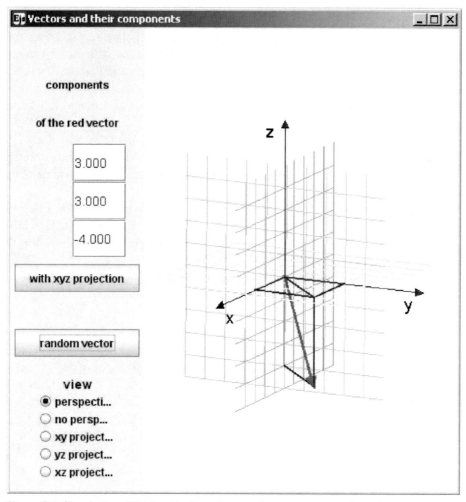

Figure 8.1. Simulation. 3D visualization of vectors in space: presentation of the components. The orientation of the projection can be adjusted with the mouse. The components of the red vectors can either be entered as numbers or created by a random number generator. The option boxes allow the selection of different projections.

corresponding to the vector format of the formula editor *MathType*, bold letters in the font *Times New Roman*.

The absolute value of the vector (the length of the arrow) is symbolized by surrounding the vector by |-signs. This is analogous to the notation for the absolute value of complex numbers, but the notions of absolute value are not quite identical. The length of a vector is independent of its position relative to the origin of a coordinate system, while the absolute value of a complex number is always calculated from the origin. This difference falls away if one writes a vector that starts from a point x_1, y_1, z_1 and leads to a point x_2, y_2, z_2 as the difference of two zero-point vectors, i.e. $x_2 - x_1, y_2 - y_1, z_2 - z_1$.

The interactive 3D simulation in Figure 8.1 trains the spatial perception of vectors. Pressing the *Random vector* button generates a zero-point vector with random integer coordinates (minimum -5, maximum 5) and represents it as a red arrow, embedded into a spatial tripod and supplemented by projections on the various coordinate planes, which can be switched on or off. It is advisable to pull this simulation to full screen size.

The coordinates of the vector are shown as projections onto the planes $x = 0$, $y = 0$ and $z = 0$ and are given in three coordinate fields. In these fields different *arbitrary* coordinates can be entered in order to study the effect on the position of the vector.

Alternatively the tip of the vector can be pulled with the mouse and the effect on the coordinates can be studied in two planes. The 3D projection can be also be rotated in space with the mouse. In addition, certain well-defined projections can be directly obtained via option switches.

Instructions for experiments can be found on the description pages of the simulation.

8.3 Basic operations of vector algebra

8.3.1 Multiplication by a constant

For vectors one can define the multiplication by a constant k and the addition of vectors in a meaningful way. For multiplication by a constant this is immediately obvious:

$$k\mathbf{a} = k \begin{pmatrix} a_1 \\ a_2 \\ a_3 \end{pmatrix} = \begin{pmatrix} ka_1 \\ ka_2 \\ ka_3 \end{pmatrix},$$

$$k|\mathbf{a}| = \sqrt{(ka_1)^2 + (ka_2)^2 + (ka_3)^3} = k\sqrt{a_1^2 + a_2^2 + a_3^3} = k|\mathbf{a}|. \quad \text{q.e.d.}$$

8.3.2 Addition and subtraction of vectors

For addition and subtraction the following definitions apply:

$$\mathbf{a} + \mathbf{b} = \begin{pmatrix} a_1 \\ a_2 \\ a_3 \end{pmatrix} + \begin{pmatrix} b_1 \\ b_2 \\ b_3 \end{pmatrix} = \begin{pmatrix} a_1 + b_1 \\ a_2 + b_2 \\ a_3 + b_3 \end{pmatrix}$$

$$\mathbf{a} - \mathbf{b} = \begin{pmatrix} a_1 \\ a_2 \\ a_3 \end{pmatrix} - \begin{pmatrix} b_1 \\ b_2 \\ b_3 \end{pmatrix} = \begin{pmatrix} a_1 - b_1 \\ a_2 - b_2 \\ a_3 - b_3 \end{pmatrix} = - \begin{pmatrix} b_1 - a_1 \\ b_2 - a_2 \\ b_3 - a_3 \end{pmatrix} = -(\mathbf{b} - \mathbf{a}).$$

The rules for multiplication by a constant and for addition and subtraction formally correspond to those for complex numbers, which one can write in analogy to the vector notation above as a matrix with one column/line and two lines/columns. Thus these vector operations are also commutative, associative and distributive, i.e. the sequence of the vectors does not matter.

Vectors, however, do *not* constitute an extension of the complex number space to higher dimensions. Vectors cannot be multiplied with each other according to the rules of complex numbers, and the division of one vector by another one cannot be defined.

In the following, we will define two different kinds of *multiplications between* vectors. These are operations that do not have an analogue in the space of the real or complex numbers. They are, rather, newly introduced for reasons of expediency. It is somewhat unfortunate that the term *multiplication* has been used. Experts also feel this way, which can be seen from the fact that what used to be referred to as the *scalar product* in earlier times, is nowadays preferably called the *inner product*, and what used to be called the *vector product* is nowadays referred to as the *outer product*. This is, however, only a semantic problem, as soon as one understands the specifics.

8.3.3 Scalar product, inner product

For the vector addition we assume that both vectors are similar quantities, that is, for example, that they represent two forces or two distances. It would not make sense to add a force to a distance, although they are both represented by vectors.

However, in physics one would like to combine two vectors of different types with each other. Force and distance are suitable examples. We define *work = force times distance*, where the quantities of force and distance enter with the length of the corresponding vectors. For this easy formula we assume that the directions of the force and distance vectors coincide. If, however, the force acts in another direction, for example at a right angle to the direction of movement, the force acting at a right angle does not perform any work. The interplay between the force- and distance vectors thus depends not only on the absolute value of the two vectors but also on the angle between them.

The corresponding combination of two vectors $\mathbf{a}$ and $\mathbf{b}$ is denoted as the *scalar product* or *inner product* and is defined by

$$\mathbf{a} \bullet \mathbf{b} = |\mathbf{a}|\,|\mathbf{b}|\cos(\mathbf{a},\mathbf{b}) = \begin{pmatrix} a_1 \\ a_2 \\ a_3 \end{pmatrix} \bullet \begin{pmatrix} b_1 \\ b_2 \\ b_3 \end{pmatrix}$$

$$= a_1 b_1 + a_2 b_2 + a_3 b_3 = b_1 a_1 + b_2 a_2 + b_3 a_3 = \mathbf{b} \bullet \mathbf{a},$$

where $(\mathbf{a}, \mathbf{b})$ is used as a sign for the angle between $\mathbf{a}$ and $\mathbf{b}$. For the combination sign, a dot is used and the combination is read as *a dot b*.

The inner product is a maximum if both vectors are parallel ($\cos(0) = \cos(\pi) = 1$) and is equal to zero if they are orthogonal to each other ($\cos(\pi/2) = \cos(\frac{3\pi}{2} = 0)$). It is a *number*, a *scalar*,[16] not a vector. This product is commutative, i.e. it does not matter which vector appears first in the product. The resulting number is equal to the length of one of the vectors multiplied by the projection of the other vector onto it.

8.3.4   Vector product, outer product

A second well defined way to combine two vectors of different types *defines* a *vector* as the result of a multiplication. Its direction is orthogonal to both input vectors and therefore also on the plane defined by the two vectors. Its absolute value is $|\mathbf{a}| \times |\mathbf{b}|\sin(\mathbf{a},\mathbf{b})$. The product is a maximum if both vectors are orthogonal to each other. An example from physics is the deflecting force on a moving charge in a magnetic field.

For this *outer product* or *vector product* the following definitions apply:

$$\mathbf{c} = \mathbf{a} \times \mathbf{b} = \begin{pmatrix} a_1 \\ a_2 \\ a_3 \end{pmatrix} \times \begin{pmatrix} b_1 \\ b_2 \\ b_3 \end{pmatrix} = \begin{pmatrix} a_2 b_3 - a_3 b_2 \\ a_3 b_1 - a_1 b_3 \\ a_1 b_2 - a_2 b_1 \end{pmatrix} = - \begin{pmatrix} a_3 b_2 - a_2 b_3 \\ a_1 b_3 - a_3 b_1 \\ a_2 b_1 - a_1 b_2 \end{pmatrix} = -\mathbf{b} \times \mathbf{a},$$

$|\mathbf{c}| = |\mathbf{a}|\,|\mathbf{b}|\sin(\mathbf{a},\mathbf{b})$.

This initially slightly confusing formula for the resulting vector can easily be analyzed mnemonically: in the first component the first coordinate of the input vector does not appear, and in its negative term the indices are simply exchanged. For the second and third component the indices are cyclically changed.

$$\mathbf{a} \times \mathbf{b} \text{ is read as } a\ cross\ b \text{ (versus } \mathbf{a} \bullet \mathbf{b} \text{ as } a\ dot\ b).$$

The vector product is *not* commutative; it does indeed depend on the sequence of the vectors. Swapping the sequence changes the sign.

Since $\mathbf{a} \times \mathbf{b}$ is a vector one can multiply this resulting vector with a third vector $\mathbf{c}$ both in the inner as well as the outer sense. Then we have:

$$(\mathbf{a} \times \mathbf{b}) \bullet \mathbf{c} \text{ is a scalar, } (\mathbf{a} \times \mathbf{b}) \times \mathbf{c} \text{ is a vector.}$$

---

16 In mathematics a scalar is a quantity which is fully determined by specifying a number.

8.4 Visualization of the basic operations for vectors

The basic operations for vectors, i.e. *summation*, *subtraction*, and *inner* and *outer product*, which have been sketched above, are visualized in the following interactive 3D simulation, Figure 8.2. This simulation starts by creating two randomly orientated position vectors (zero-point vectors) **a** and **b** of length 1, which are embedded in a transparent sphere of radius 1. In the figure the summation of these two vectors is shown.

The orientation of the axes in space can be *adjusted* with the mouse and will be perceived as rotation of the sphere. Every activation of the button *new vectors* creates a new pair of vectors. The coordinates of these vectors are shown on the right.

Using the option switches on the left, different well-defined viewing projections can be selected.

Next to the vector switch, the angle between the vectors, the product of their absolute values (here always 1 because of normalization), the scalar product and the absolute value of the vector product, are displayed.

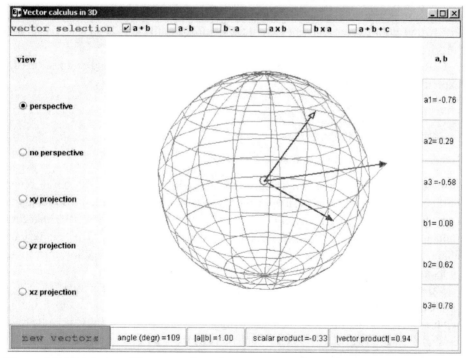

Figure 8.2. Simulation. 3D presentation of vectors, their sum, difference, outer product and multiple sum. The vector coordinates are shown on the right, the angle between the vectors, their inner product and the absolute value of the outer product on the bottom. On the left, different projections can be chosen. The Figure shows the perspective representation of the vector sum **a** + **b**. Pushing the button *new vectors* creates randomly orientated vectors.

With the option switches on top the different vector operations are visualized and superpositions are possible.

For *addition* and *subtraction* the input vectors are complemented by lines related through parallel translation. This visualizes the construction of the red result vector from the parallelograms.

$\mathbf{a} \times \mathbf{b}$ creates the *vector product* **a** cross **b** and displays it as black arrow. If the sphere is rotated in such a way that the plane defined by the two input vectors lies in the figure plane, then these vectors just touch the equator of the sphere and one looks along the direction of the resulting vector. This demonstrates the orthogonal direction. If one moves with the right-hand side from **a** via **b** to the vector product one completes a right handed screw.

Performing the same experiment with $\mathbf{b} \times \mathbf{a}$, one completes a left handed screw. This is the meaning of the non-commutativity of the vector product: the direction of the vector product $\mathbf{b} \times \mathbf{a}$ is opposite to that of $\mathbf{a} \times \mathbf{b}$, thus we have $\mathbf{b} \times \mathbf{a} = -\mathbf{a} \times \mathbf{b}$. If one displays $\mathbf{a} \times \mathbf{b}$ next to $\mathbf{b} \times \mathbf{a}$, one sees that both vectors have the same length, but point in opposite directions.

Finally $\mathbf{a} + \mathbf{b} + \mathbf{c}$ creates three random vectors and their red sum vector. If one activates $\mathbf{a} + \mathbf{b}$ in addition, one recognizes the partial construction of the sum of the first two vectors and one can implement the completion to the total sum vector in ones imagination.

In the description of the simulation you find further details and suggestions for experiments.

8.5 Fields

8.5.1 Scalar fields and vector fields

In practical situations, the simple case that a single force vector acts on an object will occur relatively seldom. An approximate example of this would be the collision of two bodies in outer space sufficiently far away from other bodies, such that their influence can be neglected. One could then consider one of the bodies to be at rest and characterize the other one with a vector whose absolute value and direction correspond to its momentum $\mathbf{p} = m\mathbf{v}$.

Much more common is the situation where there are influences at every $\mathbf{r} = (x, y, z)$ in space on the object of interest. They can either be described by vectors, having length and direction, for example the gravitational force in the vicinity of a planet, or by scalars, which have no direction, such as the density of an atmosphere or the temperature. Both quantities, force and density, influence the movement of a test body in the vicinity of the planet. The gravitation has the effect of a directed acceleration, while the density causes a deceleration independent of the direction of movement but dependent on the position.

In the first case we call it a vector field and in the second, a scalar field. In both cases the characteristic quantities depend on the space coordinates, thus for the vector field the absolute value and direction, and for the scalar field, the value. For the case of a non-stationary field they also depend on the time.

We visualize both distributions in such a way that you will have the opportunity to edit formulas for the position dependence of absolute value and direction, or to design them yourself. This will give you a feeling for the characteristics of typical fields.

8.5.2 Visualization possibilities for scalar and vector fields

To visualize a scalar field in all generality one would need four variables, three for the position coordinates and one for the position dependent scalar itself. This information can obviously not be represented with a 3D simulation that is projected on a plane. In addition some fields will change with time as a fifth variable. Thus one has to work with certain restrictions for the visualization.

For stationary problems, the time does not play a role as variable.

For problems with rotational symmetry, for example the gas density distribution ρ around a planet with rotational symmetry around an axis, one can restrict the presentation to a cross section through the center of the planet at a right angle to this axis and plot the gas density ρ as third coordinate over the cross section xy. Thus one obtains a 3D surface in the space $xy\rho$. The field distribution in space then shows rotational symmetry with respect to the distribution on the cross section.

A second possibility for this example would be to ask where the curves of equal density are located and to create a family of such curves as a *contour plot*. This task can be solved computationally by intersecting the planes $\rho = $ constant for values on an equidistant ρ-grid with the 3D surface $xy\rho$ and finding the intersection curves. This contour plot then has the familiar appearance of a geographical contour lines display.

In the general case one would have to produce a family of such presentations for the different values of those variables that have been neglected so far. Fortunately, however, the cases of practical interest are mostly stationary and possess high symmetry, such that the methods described above can visualize the important characteristics quite well.

For vector fields, one has to show in addition the direction of the vectors localized in space and their absolute value. This requires further restrictions for the visualization.

One is mostly interested in the general structure of the field, which can be shown by putting arrows on a regular grid that show directions and absolute values at the respective positions. If one only wants to show the direction of the vectors, one can use the same length for all arrows, which makes the presentation clearer. To indicate the absolute value one can then use different shades of color.

For the presentation of a three-dimensional vector field one can stack several such cross sections over each other. As a static picture such a 3D projection is often quite confusing. However, if one moves the projection direction interactively, either with the mouse or automatically around an axis, one obtains a rather good idea of the distribution.

All these tools are provided by common numerical programs and we will show examples for these in the following.

8.5.3 Basic formalism of vector analysis

Pure scalar fields without relation to a vector field, (the density distribution is such an example), are not very interesting. Much more interesting are scalar fields, from which vector fields can be deduced. This simplifies their description greatly and shows them as originating from one position dependent parameter only.

We refer to a scalar *potential field* P, if the components of a vector field $\mathbf{V}$ are obtained by taking partial derivatives with respect to x, y, z of P,[17] i.e. via differentiation after one variable at a time, while the other variables are considered as constant. The underlying questions is, then, how the scalar value changes if one moves from a space point $\mathbf{r} = (x, y, z)$ to a neighboring space point $\mathbf{r} + d\mathbf{r} = (z + dx, y + dy, z + dz)$. Thus one can take, for each variable, i.e. partially, the first term of the Taylor expansion, if $d\mathbf{r}$ is small enough. One then obtains:

$$dP = P(x + dx, y + dy, z + dz) - P(x, y, z)$$

$$= \frac{\partial P}{\partial x}dx + \frac{\partial P}{\partial y}dy + \frac{\partial P}{\partial z}dz = \left(\frac{\partial P}{\partial x}, \frac{\partial P}{\partial y}, \frac{\partial P}{\partial z}\right)\begin{pmatrix} dx \\ dy \\ dz \end{pmatrix} = \mathbf{grad}\, P \bullet d\mathbf{r}.$$

The vector called **grad** P denotes the change of the scalar P in the three spatial directions. For a given space point, its direction depends on the change of potential in the three directions; it points in the direction of maximum change. Its absolute value depends on the absolute values of these changes.

$$\mathbf{V} = \mathbf{grad}\, P = \begin{pmatrix} \frac{\partial P}{\partial x} \\ \frac{\partial P}{\partial y} \\ \frac{\partial P}{\partial z} \end{pmatrix}; \quad |\mathbf{V}| = \sqrt{\left(\frac{\partial P}{\partial x}\right)^2 + \left(\frac{\partial P}{\partial y}\right)^2 + \left(\frac{\partial P}{\partial z}\right)^2}.$$

As shorthand for the partial differentiation with respect to all three coordinates, that is applied to the scalar potential, one uses the symbol *nabla* (∇), an overturned Greek letter Δ. This symbol reminds one of the form of an antique harp ($\nu\alpha\beta\lambda\alpha$ in Greek, *nablium* in Latin). *Nabla* symbolizes a vector operator, which is therefore written as a matrix with one column or one line. To stress the vector character of this operator,

---

17 The symbol for the partial derivative of a quantity $A(x, y, z)$ after the variable x is $\frac{\partial A}{\partial x}$

one usually puts an arrow on top of it.

$$\vec{\nabla} = \begin{pmatrix} \frac{\partial}{\partial x} \\ \frac{\partial}{\partial y} \\ \frac{\partial}{\partial z} \end{pmatrix}; \quad \mathbf{V} = \vec{\nabla} P = \begin{pmatrix} \frac{\partial}{\partial x} \\ \frac{\partial}{\partial y} \\ \frac{\partial}{\partial z} \end{pmatrix} P = \begin{pmatrix} \frac{\partial P}{\partial x} \\ \frac{\partial P}{\partial y} \\ \frac{\partial P}{\partial z} \end{pmatrix}; \quad \vec{\nabla} P = \mathbf{grad}\ P.$$

Using the nabla notation has the advantage that it also allows a unified notation for other differential operators, for which different symbols are traditionally used, which hide their common origin. For example the vector field characterized by $\vec{\nabla} P$ is traditionally denoted by **grad** P (gradient of P) and referred to as gradient field, because if characterizes the steepness of the potential field.

We will now show some further applications of the *nabla* symbol and its traditional synonyms. In the first two examples, the operator will not be applied to a *scalar field* but to a *vector field*. In analogy to the gradient of a scalar field we now deal with the change of a vector field from a space point $\mathbf{r}$ to a neighboring point $\mathbf{r} + d\mathbf{r}$.

$$\vec{\nabla} \bullet \mathbf{a} = \begin{pmatrix} \frac{\partial}{\partial x} \\ \frac{\partial}{\partial y} \\ \frac{\partial}{\partial z} \end{pmatrix} \bullet \begin{pmatrix} a_x \\ a_y \\ a_z \end{pmatrix} = \frac{\partial a_x}{\partial x} + \frac{\partial a_y}{\partial y} + \frac{\partial a_z}{\partial z};$$

$$\vec{\nabla} \bullet \mathbf{a} = \mathrm{div}\ \mathbf{a}\ divergence\ (scalar\ field)\ of\ \mathbf{a}$$

$$\vec{\nabla} \times \mathbf{a} = \begin{pmatrix} \frac{\partial}{\partial x} \\ \frac{\partial}{\partial y} \\ \frac{\partial}{\partial z} \end{pmatrix} \times \begin{pmatrix} a_x \\ a_y \\ a_z \end{pmatrix} = \begin{pmatrix} \frac{\partial a_z}{\partial y} - \frac{\partial a_y}{\partial z} \\ \frac{\partial a_x}{\partial z} - \frac{\partial a_z}{\partial x} \\ \frac{\partial a_y}{\partial x} - \frac{\partial a_x}{\partial y} \end{pmatrix};$$

$$\vec{\nabla} \times \mathbf{a} = \mathbf{curl}\ \mathbf{a}\ curl\ (vector\ field)\ of\ \mathbf{a}$$

$$\vec{\nabla}^2 = \begin{pmatrix} \frac{\partial}{\partial x} \\ \frac{\partial}{\partial y} \\ \frac{\partial}{\partial z} \end{pmatrix} \bullet \begin{pmatrix} \frac{\partial}{\partial x} \\ \frac{\partial}{\partial y} \\ \frac{\partial}{\partial z} \end{pmatrix} = \frac{\partial^2}{\partial x^2} + \frac{\partial^2}{\partial y^2} + \frac{\partial^2}{\partial z^2}; \quad \vec{\nabla}^2 = \Delta\ Laplace\ operator$$

$$\vec{\nabla}^2 P = \Delta P = \mathrm{div}\ \mathbf{grad}\ P; \quad Laplace\ P\ \text{(referred to as the "Laplacian",}$$
$$\text{which is a scalar field).}$$

The meanings of the symbols and operations are, in short, as follows:

The **divergence** of a vector field in Cartesian coordinates is obtained computationally as the (symbolic) *scalar* multiplication of the nabla operator with the vector and therefore it is a scalar field. It describes the *source strength* of the vector field. Where it does not vanish, field lines either originate or converge.

An example from the Maxwell equations: div $\mathbf{D} = \rho$; the charges ρ are the sources of the electrical vector field $\mathbf{D}$.

The **curl** of a vector field is obtained computationally as (symbolic) *vectorial* multiplication of the *nabla* operator with the vector field, thus at every space point, it is a vector. The curl of a vector field describes the *vorticity* of a vector field, which has closed field lines, unless **curl a** $= 0$ everywhere.

Another example from the Maxwell equations: **curl H** $=$ **j**; the current density **j** determines the vector field **H** that is orthogonal to the current density and has closed field lines everywhere.

The **Laplace operator** is obtained via (symbolic) *scalar* multiplication of the nabla operator with itself and therefore yields a scalar field.

An example for its application: under the assumption that the electrical field strength is the gradient field of an electrostatic potential, i.e. $\mathbf{E} = -\mathbf{grad}\ \Phi$, we obtain from one of the Maxwell equations, namely div $\mathbf{D} = \rho$, the Poisson equation $\Delta \Phi = -\rho/\varepsilon_0$. Using this equation the electrostatic potential due to a given charge density ρ can be calculated, and from this the electrical vector field. In a portion of space without charges the **potential equation** $\Delta \Phi = 0$ applies.

Between the different operators, the following general relations apply:

For every scalar field V we have: **curl grad** $V = \vec{\nabla} \times \vec{\nabla} V = 0$, i.e. for a gradient field the (local) curl is zero, there are no vortices.

div **curl a** $= 0$. The (local) divergence of the curl field of a vector field is zero, because a pure vortex field does not have any sources.

8.5.4 Potential fields of point sources as 3D surfaces

Particularly elementary, simple, and at the same time important, are the potential fields that are caused by point sources in space. They describe both the gravitational attraction between masses m_i as well as the attraction or repulsion between charges e_i, which can be positive or negative. The common property of these forces is that, with growing distance, the effect of the point source is spread over the surface of a sphere and therefore decreases like $1/4\pi r^2$. The potential field then has, apart from an additive constant as integral of the vector field in polar coordinates, the form:

$$-\frac{m_i}{r}, \quad \text{or} \quad -\frac{e_i}{r}.$$

The effect on a similar object increases with decreasing distance, because $1/r$ becomes larger, when the distance becomes smaller. The minus sign has the effect that the force $\mathbf{F} = -\mathbf{grad}\ P = (m/r^2)\mathbf{r_0}$ is positive if m_i is positive. In the case of gravitation, this is always the case. In the electrostatic example, the increasing repulsion of equally charged objects turns into an increasing attraction, if they are of opposite charge.

The following interactive simulation of scalar fields shows as examples:

- the potential field of a point source;
- the potential field of two point sources of equal sign at a distance of r, with adjustable mass ratio (charge ratio) b;
- the potential field of three symmetrically located point sources of equal sign at a pairwise distance r, with adjustable mass ratios $b = m_2/m_1, c = m_3/m_2$ or charge ratios $b = e_2/e_1, c = e_3/e_2$;
- the potential field of a dipole consisting of a negative and equally large positive charge at a distance of r;
- the potential field of a quadrupole consisting of two dipoles arranged symmetrically at a distance r.

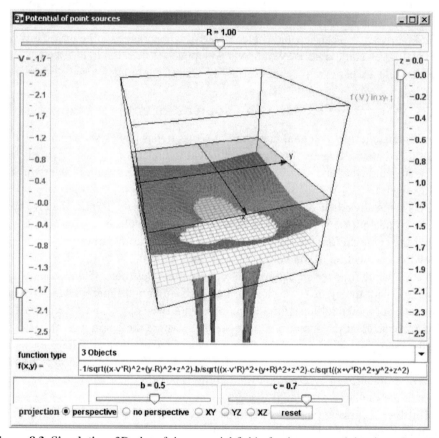

Figure 8.3. Simulation. 3D plot of the potential field of point sources lying in a plane. The figure shows the field of three similar point sources. With the right slider the mass or charge can be fixed, with the left slider the yellow intersection plane can be shifted and with the bottom sliders b, c the mass or charge ratios can be set. In addition, different projections of the 3D field can be chosen.

The first object is normalized to 1; the distance r can be changed continuously with a slider.

The potential distribution (the value P can be chosen with the left slider), is calculated for planes in an adjustable distance z to the xy-plane, which can be set with the right slider. The potential distribution in the z-plane is shown by the intersection curve between the potential surface and the z-plane.

The fields diverge at the respective point sources, since $\lim_{r \to 0}(\frac{1}{r}) = \infty$. In the simulation, this is prevented by excluding the plane $z = 0$. For a realistic field distribution one would have to work with extended charged objects instead of with point sources.

This simulation provides many possibilities, whose detailed description cannot be given here. The formulas are editable, such that, in addition to the given fields, additional fields can be calculated. In the description pages we discuss this further and useful experiments are proposed.

In Figure 8.3 you can see the whole interactive appearance for the three body problem with three equal objects. We show a potential cross section in the plane $z = 0$ in the "far field", where the potential surface has not yet split into partial surfaces.

8.5.5 Potential fields of point sources as contour diagrams

The visualization of potential curves in cross sectional planes shown above is very flexible. However, it takes some careful thought to understand what is actually being calculated and what one sees. The extensive details in the description pages will assist in this regard.

This is the advantage of presentation as a *contour diagram*. It immediately shows a family of potential curves of equal potential distance in a plane.

For the computation, the same algorithm as above is used. However, now we only show the intersection of the yellow plane from Figure 8.3 with the potential surface and at the same time we show a number of potential lines (here 35, they cannot all be separated with the eye). The following interactive diagram Figure 8.4a shows in the xy-plane equipotential lines for a large mass with two smaller masses in its vicinity. One recognizes at the same time the *near field*, where the equipotential lines encircle the individual objects, and the *far field*, where the equipotential lines encircle all objects, as well as the *neutral* points, with **grad** $P = 0$, in which an object without its own momentum would not know where to turn (the force as gradient acts in the direction of the largest potential change).

The three following static pictures show next to each other the three-body potential of three equal bodies in a plane with distance $z = 0.42$ to the xy-plane, as well as a dipole and a quadrupole field, which have a very instructive appearance in this presentation. One must, however, take into account that the equipotential lines in the individual z-planes do not represent identical potentials. The 35 potential lines are

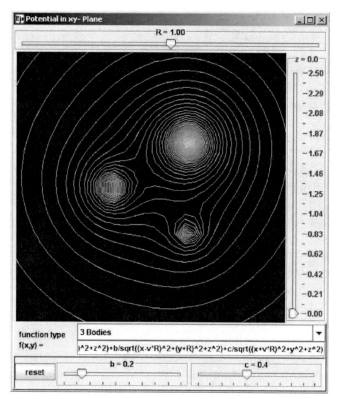

Figure 8.4a. Simulation. Contour representation of a potential field, in the picture the field of a point source with two smaller satellites (mass ratio 0.2) is shown. The scanning plane can be moved with the right slider. Thirty-five potential lines are calculated. That these appear to have corners at small distances is an artifact of the calculation.

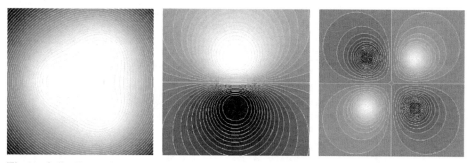

Figure 8.4b. Examples: contour diagrams for three point masses, for a dipole of a positive and a negative charge and a quadrupole formed from two neighboring dipoles.

determined separately for each layer. Thus one obtains a qualitative picture, while the 3D presentation is quantitative.

By moving the z-plane, one quickly obtains in this simulation an idea about the spatial distribution of the potential.

8.5.6 Plane vector fields

In reality, vector fields have many components, derivatives and variables: three for the position, and three for the components of the direction, the absolute value and the time. For numerical calculation this does not really pose a problem; one simply needs to do the calculation with the required number of dimensions.

However, one has to accept many limitations for the visualization, since the desired relationships have to be shown as a projection on a plane. It becomes relatively easy if one assumes that the vector field is in a stationary state (no time dependence of the direction or absolute value) and if we restrict ourselves to vectors in a plane, as we will do in this subsection.

The local distribution of the vector *direction* can be shown quite clearly with arrows, whose origins are placed on a regular grid in the plane. The pictorial presentation of the *absolute value* of the vector is, however, less convincing. If one chooses the length of the arrows to show this, the arrangement easily becomes unclear, since the dependence on the position can be quite strong, for example for a quadratic dependence from the distance to the source. If one chooses different shades of color, the achievable range is small and the presentation is only of a qualitative nature.

We have chosen a uniform arrow length. Color shading gives a qualitative hint about the absolute value of the vector field. For the quantitative presentation of the absolute value of the vector, we use the velocity of a test object that moves in the vector field (a_x, a_x). Thus the time is used as another dimension of the presentation.

The red *test object* moves from the start along the field lines with a velocity that is determined from the components of the vector field a_x, a_y according to very simple, coupled ordinary differential equations (see Chapter 9):

$$v_x = \frac{da_x}{dt}; \quad v_y = \frac{da_y}{dt}.$$

In two dimensions, particularly, the formula for the curl becomes clearer:

$$\mathbf{curl\, a} = \vec{\nabla} \times \mathbf{a} = \begin{pmatrix} \frac{\partial}{\partial x} \\ \frac{\partial}{\partial y} \end{pmatrix} \times \begin{pmatrix} a_x \\ a_y \end{pmatrix} = \begin{pmatrix} 0 \\ 0 \\ \frac{\partial a_y}{\partial x} - \frac{\partial a_x}{\partial y} \end{pmatrix};$$

$$\mathrm{div\, a} = \vec{\nabla} \cdot \mathbf{a} = \begin{pmatrix} \frac{\partial}{\partial x} \\ \frac{\partial}{\partial y} \end{pmatrix} \cdot \begin{pmatrix} a_x \\ a_y \end{pmatrix} = \frac{\partial a_x}{\partial x} + \frac{\partial a_y}{\partial y}.$$

The curl has only *one* component, namely in the z-direction, since it needs to be orthogonal to the xy-plane of the vector field. The source strength only depends on the changes in the x-and y-directions.

In general, the components of the vectors will be functions of both variables x and y, $a_x = a_x(x, y)$; $a_y = a_y(x, y)$, such that the scalar *divergence* and the absolute value of the *curl* depend on the position. The direction of the curl is for a plane field always the orthogonal axis, normally called the z-axis.

In two dimensions, the curl can be calculated easily for given formulas of the components. Remember that one treats the respective second variable as constant for the partial differentiation with respect to the first one.

Vortices in the vector field can be easily recognized in the chosen presentation.

A vector field is *vortex free* if its curl is vanishes everywhere:

$$\textbf{curl a} = 0 \quad \text{for} \quad \frac{\partial a_y}{\partial x} - \frac{\partial a_x}{\partial y} = 0 \rightarrow \frac{\partial a_y}{\partial x} = \frac{\partial a_x}{\partial y}.$$

This is for example the case, if $a_x = a_x(x)$; $a_y = a_y(y)$ i.e. if the components only depend on their own coordinate. In this case, the partial derivatives vanish identically (further details are given in the simulation description).

Sources in a vector field can be recognized visually by sequences of vectors, i.e. *field lines* that start or end at them. A field is free of sources and sinks (negative sources), if the divergence vanishes *everywhere*.

$$\text{div a} = 0 \quad \text{for} \quad \frac{\partial a_x}{\partial x} + \frac{\partial a_y}{\partial y} = 0 \rightarrow \frac{\partial a_x}{\partial x} = -\frac{\partial a_y}{\partial y}.$$

This is, for example, the case when the vector components are independent of the coordinates, i.e. if the derivatives vanish identically.

In other cases, one needs to examine critically if the formally satisfied condition provides an answer that make sense *in all points* of the vector field. This is, for example, not the case if the limiting process when calculating the derivative results in an undetermined expression $(0/0)$. A secure statement is obtained if one surrounds the suspected source with a circle (a sphere in three dimensions) and sums up the number of field lines that cross the curve while taking the signs into account. If the number of field lines entering the circle is the same as the number of field lines leaving it, the corresponding point is free of sources. This statement becomes exact when integrating over the volume and taking the limit of vanishing radius.

Figure 8.5, which shows at the start a vector field with two vortices, leads to the interactive simulation. The test object, which is initially at rest (initial velocity 0), can be *moved* to an arbitrary position in the field using the mouse, prior to the time simulation, in order to investigate the total field in detail.

With the *selection field* one can choose one of a number of typical fields. The formulas for the components, as well as the respective divergence and curl, are then given

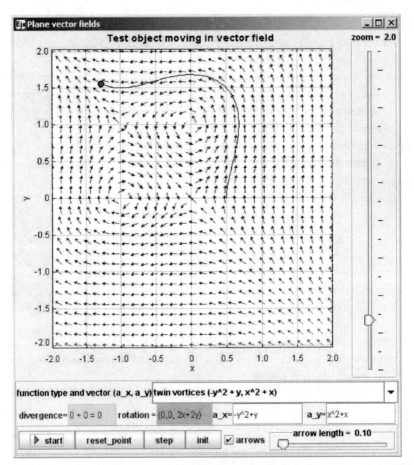

Figure 8.5. Simulation. Animated movement of a test body in a given vector field. The topmost, non-editable, text field shows preset field types and their vector components a_x, a_y. The components can be edited separately in the small text fields, such that arbitrary fields can be created. The right slider controls the zoom factor, the lower slider, the arrow length that is constant over the whole plane. For the preset fields, divergence and rotation are displayed.

in text fields. The formulas field can be edited, so that arbitrary component formulas can be entered to study the corresponding fields.

The scale slider *zoom* on the right allows the investigation of the field on larger or smaller scales. This variation possibility is important, because the number of arrows shown is constant for clarity, but at a larger scale, details such as vortices and sources can be lost.

The *arrow length*, which is constant for the whole plane, can be adjusted with the second slider.

Further details and suggestions for experiments are given in the description pages of the simulation.

8.5.7 3D field due to point charges

Figure 8.6 shows the vector field of a quadrupole. It leads to the simulation of the general electric vector field of point charges. The distribution of its direction is visualized through a periodic grid of arrows having constant length, which show the local direction of the electric field strength vector in every point. The vector length is adjustable, but constant everywhere within the picture.

The absolute value of the field strength is indicated by color shading. In addition, a threshold value for the lowest absolute value for which vectors are shown can be selected. This provides a field strength dependent envelope surface for the whole field.

The opaque yellow plane can be shifted parallel to the z-axis, such that a spatial cross section through the vector field is shown.

The space orientation of the presentation can be adjusted with the mouse; in addition, defined projections can be selected.

The number of objects can be chosen freely, and one can switch between particles of the same or opposite polarity. This shows the considerable difference in fields between multipoles of opposite polarity and particle configurations of uniform polarity, most distinctly recognized in the far field.

In the initial state, all particles are positioned uniformly on a circle around the origin in the xy-plane. They can be individually *moved* with the mouse, such that arbitrary configurations are possible.

A convincing visualization of a situation that depends on so many parameters via projection to the observation plane requires a careful coordination of point distance, arrow length, threshold level and observation angle. The spatial impression becomes quite vivid if one changes the orientation of the projection slowly by pulling with the mouse.

The description contains further details and suggestions for experiments.

8.5.8 3D movement of a point charge in a homogeneous electromagnetic field

The movement of a charge in an electric field is quite easy to understand. It follows the direction of the electric field and the charge is accelerated proportionally to the absolute value of the electric field vector.

The movement in a magnetic field is much more complicated. In this case, the vector product of magnetic field and velocity determines the acceleration of the charged test mass. Thus the charge is deflected at a right angle both to the magnetic field and the direction of its velocity, and the strength of the deflection depends on the angle between magnetic field and current direction of movement, namely $\mathbf{F} \sim \mathbf{v} \times \mathbf{B}$. The effect of this force is that the orbit moves in spirals around the direction of the magnetic field lines.

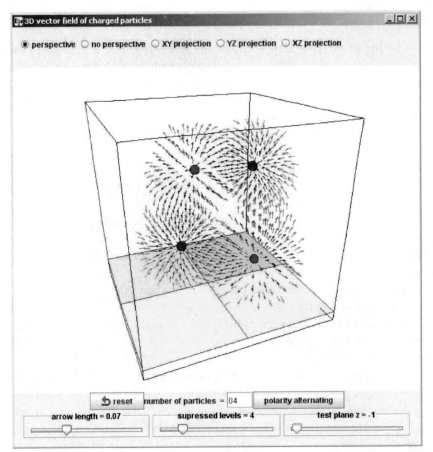

Figure 8.6. Simulation. 3D field of point sources that are located at arbitrary positions; in the figure a quadrupole consisting of two positive and and two negative point charges. Using the sliders the arrow length, the threshold of the displayed field strength level, and the position of the yellow scanning plane, can be adjusted. The number of charges is entered by hand in the number field while the switch determines whether the charges all have the same or alternating opposite signs. On the top, different projections can be selected. All individual particles, as well as the 3D projection, can be moved or adjusted with the mouse.

    Through the combined effect of magnetic and electric fields, the accelerations are added and very different movement patterns can come about. We want to visualize this for the simple example of a homogeneous field, for which the electric and magnetic field are constant in absolute value and direction in the whole space.

    The interactive Figure 8.7 shows after opening the movement of a charge, whose initial velocity vector has components in the positive y-direction and negative z-direction, and which is subject to the accelerating effect of the green electric field vector and the "rolling up" effect of the red magnetic field. The spiral drawn in magenta

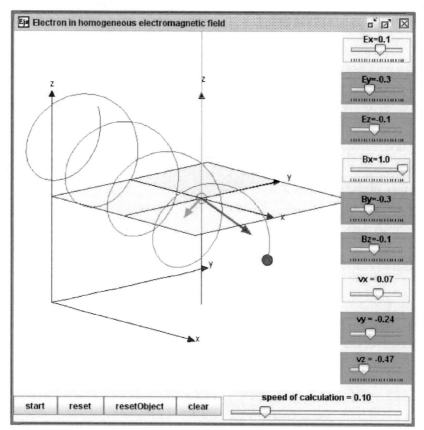

Figure 8.7. Simulation. Movement of a charge with given initial velocity vector in a homogeneous electromagnetic field. The components of the homogeneous fields and of the velocity vector are shown in the sliders at the right and can be set as initial values before starting the simulation. The velocity vector for the simulation is set with the lower slider. The simulation is started with the *Start* button.

is stretched in the direction of the E field. The axis of the spiral is parallel to the red vector of the magnetic field, as becomes apparent if turning the 3D simulation appropriately.

Arbitrary initial conditions can be set with three sliders each for the homogeneous vector components of electric field, magnetic field and for the initial velocity of the point charge. After pushing the *Start* button, which triggers the calculation of the orbit using the respective differential equations, the sliders for the velocity components move according to their changes as function of time. The speed of calculation can be adjusted.

The start point of the charge can be adjusted with the mouse. *ResetObject* moves the start point to the origin, while leaving all other settings unchanged. *Reset* changes all parameters back to their default settings. *Clear* erases all orbits. Thus one can also superimpose orbits with different settings.

The simulation provides a wealth of possibilities, of which a few limiting cases are named below:

- No fields: uniform movement of the charged particle with the initial velocity and no acceleration;
- Only E field: uniformly accelerated deflection of the object;
- Only B field and velocity parallel to the magnetic field: field has no effect on the movement;
- Only B field and velocity vector orthogonal to it: circular orbit;
- B field and E field orthogonal to each other and velocity vector orthogonal to B field: spirals.

The description of the simulation contains further details and suggestions for experiments.

9 Ordinary differential equations

9.1 General considerations

In connection with the differential quotient we have introduced the notion of *differen-* ODE
tial equations and briefly described the particularly simple and important differential
equations for trigonometric and exponential functions.

In this chapter we want to deal extensively with this "magic wand" of infinitesimal
calculus, which provides the key to a deeper understanding of physical relationships.

Which *concrete* meaning can be associated in ones imagination with the first and
second differential quotients y' and y'' of a function y (higher differential quotients
barely play a role).

We consider a graphical presentation of the function $y = f(x)$ in a plane coordinate system.

The first derivative $y'(x) = \frac{dy}{dx}(x)$ is the *slope* or *steepness* of the curve describing
the function at position x. It indicates how strongly y changes for a given x as a
function of x. Positive values signify an increase, negative values a decrease.

The second derivative $y''(x) = \frac{d^2y}{dx^2}(x) = \frac{dy'}{dx}(x)$ describes the *change of slope*
and thus the local *curvature*. Positive values mean an increase of the slope and
thus concave curvature, while negative values signify a decrease of the slope and thus
convex curvature.

We now want to interpret especially the *variable* x *as time* t; we thus consider
changes of the quantity y as function of time. An example would be a driving car, for
which y is the distance traveled during the time interval t: $y = f(t)$. Thus at time
$t = 0$ the position of the car is $y(0) = f(0)$.

The first derivative $y'(t) = \frac{dy}{dt}$ is then the *change* in the distance traveled per
infinitesimal time interval, measured at a certain time point t, and thus has the meaning
of the instantaneous *velocity* v of the car.

The second derivative $y''(t) = \frac{d^2y}{dt^2}(t) = \frac{dy'}{dt}$ describes the *change of velocity*
and thus the instantaneous *acceleration* a of the car. Positive acceleration means an
increase in the velocity, negative acceleration means a decrease in the velocity, i.e.
deceleration.

Thus, for the illustrative description of differential quotients (derivatives), the designations *slope* or *steepness* are equivalent, as well as the designations *curvature* and
acceleration.

We demonstrate here the strength of the predictive power of an extremely simple differential equation, the example of a driving car. In school we learn with great

effort the formula for the time dependence of the traveled distance s with v_0 as initial velocity and s_0 as initial value of s:

$$s(t) = \frac{a}{2}t^2 + v_0 t + s_0.$$

We also learn the restrictive condition, that the acceleration a must be constant for the equation to be correct at all (every child knows, however, that this does not happen in reality). The simple differential equation:

$$s'' = a$$

does not only apply in the same situation, but is also valid if the acceleration is not constant, but is an arbitrary function of time $a(t)$. To distinguish between all individual events that satisfy the differential equation, it is sufficient to know the respective initial values v_0 and s_0.

To calculate the time dependence from the differential equation is a routine job that is identical for all differential equations, and for which one can use the analytical tools of integral calculus, or which is just left to a numerical computational code.

In physics we often do not wish to calculate the values that result in a special case, but to primarily understand which causal relationships are behind a certain phenomenon. The example of distance traveled simply provides the answer: *the acceleration is important.*

This statement in its formal simplicity is also valid if we examine a "curved" three-dimensional orbit in space under the influence of different forces! For the force vector **F** that acts on the object with mass m we have:

$$\mathbf{F} = m\mathbf{a},$$

where **a** is the acceleration vector. This was one of Isaac Newton's greatest insights.

9.2 Differential equations as generators of functions

The example of the distance traveled also gives an easy answer to the question: how does one find the functions that are important in physics? How are functions defined that describe certain situations? How does one find the relationship between variable and function value that is expressed in the function, i.e. the "character" of a special function type?

Differential equations are the *parents* of the functions, and we will soon see that a single differential equation, i.e. a simple relationship, creates many related children – read functions.

As discussed in Chapter 5, *functions* describe the dependence of a quantity y on one or more other quantities, which are called *variables* of the function, or more correctly *independent variables*.

For one variable, which we call t, these functions shows y in its dependence from this single variable t. (We here choose t as symbol of the variable, since many examples will demonstrate a dependence on time.) The curve of the function can be visualized with a $y(t)$ plot. Changes are described by derivatives with respect to the single variable t.

$$\text{Function } y = y(t), \quad \text{variable: } t_1 < t < t_2,$$

$$\text{slope } y'(t) = \frac{dy}{dt}; \quad \text{curvature } y''(t) = \frac{d^2 y}{dt^2} = \frac{dy'}{dt}.$$

What happens if several variables need to be taken into account in the example of a time dependent function of two position variables, i.e $x = f(x, y, t)$, which we have visualized above? Here the partial differential quotients appear, which describe the change of the function value z when varying one of the variables. For the partial derivative of a function with respect to one of the variables, all other independent variables are treated as constants.

$$z = z(x, y, t)$$

$$\text{variable:} \quad x_1 < x < x_2; \ y_1 < y < y_2; \ t_1 < t < t_2;$$

$$\frac{\partial z}{\partial t}; \ \frac{\partial z}{\partial x}; \ \frac{\partial z}{\partial y}; \ \frac{\partial^2 z}{\partial t^2}; \ \frac{\partial^2 z}{\partial x^2}; \ \frac{\partial^2 z}{\partial y^2}; \ \frac{\partial^2 z}{\partial x \partial y}; \ \dots$$

We now go back to relationships for one independent variable and consider a simple example: we know about the exponential function, that its instantaneous growth, the *growth rate* or *slope*, is exactly the same as its function value. The growth rate is identical to the first differential quotient:

$$\text{for } y = e^t \text{ we have } y' = \frac{dy}{dt} = e^t;$$

$$y' = y.$$

This *differential equation* (relationship between the function and its derivatives) characterizes the nature of all growth functions (exponential functions) in a unique way. To fix a specific *growth function*, one only requires its initial value.

If, as above, we only deal with the differential dependence on *one* variable, we refer to an *ordinary differential equation*. *Partial differential quotients* appear when there is a dependence on more than one variable and we refer to a *partial differential equation* (see below).

There is no other function that shows the same property of the derivative as the exponential function. This applies irrespective of its "amplitude", i.e. a multiplicative factor C, because:

$$\text{for } y = Ce^t \text{ we have } y' = \frac{dy}{dt} = Ce^t = y \rightarrow y' = y.$$

In general, every *linear* differential equation is independent of multiplicative factors.

It is quite easy to formally derive the exponential function as a solution from the knowledge of the differential equation, using elementary integrals.

$$y' = y \equiv \frac{dy}{dt} = y;$$

the solution method of choice is the seperation of variables

$$\frac{dy}{y} = dt,$$

integration of left-hand side

$$\int_{y(0)}^{y} \frac{1}{y} dy = \ln y - \ln y_0 \quad \text{with } y(0) = y_0,$$

integration of right-hand side

$$\int_0^t dt = t - 0 \rightarrow \ln y = t + \ln y_0; \quad y = e^{t + \ln y_0} = y_0 e^t.$$

The *basis* exponential function $y = e^t$ is the result for the *initial value* $y_0 = 1$. From the last equation, one can see that multiplication with the initial value yields the same function as translation along the t-axis by the logarithm of the initial value.

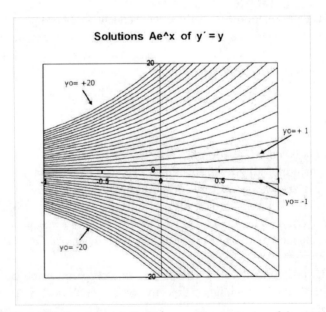

Figure 9.1. Family of solutions $y = Ae^x$ of $y' = y$. The parameter of the curves is the initial value $A = y(0)$ (intersection with the ordinate), which takes on the integer values from -20 to $+20$.

For different initial values, the differential equation describes a family of exponential functions, which are distinguished by a multiplicative factor. The diagram in Figure 9.1 shows this family of curves for positive and negative initial values between -20 and 20 with step width 1.

In general, an ordinary differential equation is defined by a *functional relationship* between a *function*, its *derivatives* and the *variable t*:

$$\text{general} \quad F(t, y, y', y'', \ldots, y^{(n)}) = 0,$$

$$\text{explicit} \quad y^{(n)} = f(t, y, y', y'', \ldots, y^{(n-1)}).$$

A differential equation is called *explicit* if the highest derivative can be expressed as a function of the lower derivatives.

The above equation for the exponential function is an ordinary explicit linear first order differential equation. The equation is:

- *ordinary*, because it only has one variable;
- *explicit*, because the derivative of highest order can be expressed as a function that does not contain itself;
- *linear*, since the function itself and all derivatives except for the highest order enter in a linear fashion;
- *of first order*, since only the first derivative appears in it.

These criteria are important for finding an analytical solution and are also important for numerical solution with limited computational power. In the case of explicit equations and important numerical methods, only previously calculated data enter the procedure to calculate a solution step by step. For implicit equations (an exotic example is $y'' \cos y'' + x^y = 0$) one has to solve for every step of the calculation an equation that already contains the results for the next step. This is, in general, not possible in closed form, but only through iteration. With sufficient computational capacity, however, this dilemma loses its importance. We have already shown above how to solve complicated equations with iterative methods.

One could, of course, also describe the exponential function via a differential equation of second, or even higher order, because we have

$$y = e^x; \ y' = e^x; \ y'' = e^x; \ \ldots$$

$$\rightarrow \text{for example} \quad y'' = y.$$

If one asks which functions satisfy this second order differential equation one realizes, maybe initially surprisingly, that it is not satisfied only by the simple exponential function but, in addition, by a multitude of functions that are related to it.

Indeed we have with A and B as constants:

For $y = Ae^t \rightarrow y' = Ae^t \rightarrow y'' = Ae^t$ $\qquad\qquad\qquad \Rightarrow y = y''(= y')$.

For $y = Ae^{-t} \rightarrow y' = -Ae^{-t} \rightarrow y'' = Ae^{-t}$ $\qquad\qquad \Rightarrow y = y''(\neq y')$.

For $y = Ae^t - Be^{-t} \rightarrow y' = Ae^t + Be^{-t} \rightarrow y'' = Ae^t - Be^{-t} \Rightarrow y = y''(\neq y')$

in particular for $A = B = 1/2$:

$$y = \cosh(t) = \frac{e^t + e^{-t}}{2} \rightarrow y' = \frac{e^t - e^{-t}}{2} \rightarrow y'' = \frac{e^t + e^{-t}}{2} \qquad \Rightarrow y = y''(\neq y')$$

$$y = \sinh(t) = \frac{e^t - e^{-t}}{2} \rightarrow y' = \frac{e^t + e^{-t}}{2} \rightarrow y'' = \frac{e^t - e^{-t}}{2} \qquad \Rightarrow y = y''(\neq y').$$

In addition to the simple exponential function with a positive exponent, this also includes exponential functions with a negative exponent and also all linear combinations of these two components, of which we have formulated $\cosh t$ and $\sinh t$ at the end of the list.

In the diagram Figure 9.2, which is *not active*, we show the families of the functions described above. First the family of exponential functions with positive and negative exponents and initial values is shown.

Figure 9.3 then shows the hyperbolic functions that are either symmetric or antisymmetric to $x = 0$ and are determined by a single initial value A:

$$A \sinh(x) = A\frac{e^x - e^{-x}}{2}; \qquad A \cosh(x) = A\frac{e^x + e^{-x}}{2}.$$

Finally Figure 9.4 shows the general solutions with two parameters A and B:

$$Ae^x - Be^{-x}; \qquad Ae^x + Be^{-x}$$

$$\text{with } A = 1, 2, 3, \ldots, 10$$

$$\text{and } B = 1, 5, 10.$$

The choice of the parameters A and B, including their signs, determines which individual function, from the abundance of functions that satisfy the differential equation $y'' = y$, is realized. One obtains *all* functions for which the curvature has the same sign and absolute value as the function value.

In this simple case it is immediately obvious that one could, instead of the second order differential equation, also use two first order differential equations:

$$
\begin{array}{ccc}
y' = y & & y' = -y \\
& \text{or} & \\
y'' = y' & & y'' = -y'.
\end{array}
$$

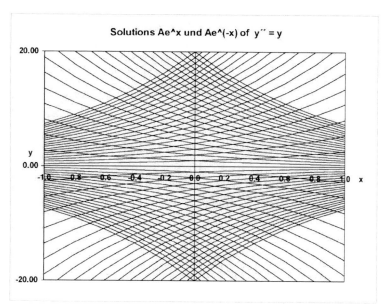

Figure 9.2. Family of solutions $y = Ae^x$ and $y = Ae^{-x}$ of $y'' = y$ for different initial values of $A = y(0)$. A covers all integer values from -20 to $+20$.

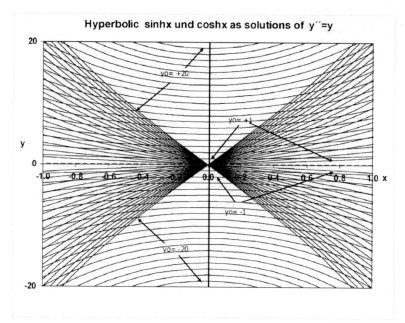

Figure 9.3. Family of solutions of $y'' = y$: the function $y(x) = A \sinh x$ goes through $(0, 0)$ and $y(x) = A \cosh x$ goes through $(0, A)$ with initial value $y(0) = A$. A changes in integer steps from -20 to $+20$.

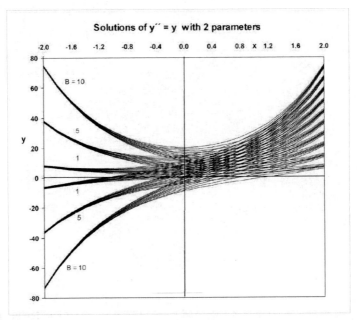

Figure 9.4. Two-parameter family of solutions of $y'' = y$. The vales of A and B determine for example the two initial values $y(0) = A + B$ and $y'(0) = A - B$.

In general, one can reduce an ordinary differential equation of nth order to a *system of n equations* of first order:

$$y^{(n)} = f(y, y', y'', \ldots, y^{(n-1)}, t) \rightarrow$$

1) $y' = y_1$

2) $y_1' = y_2$

3) $y_2' = y_3$

4) $y_3' = y_4$

$$\vdots$$

$$y^{(n)} = f(y, y_1, y_2, y_3, \ldots, y_{n-1}, t).$$

Since the differential equation is explicit, $f(\ldots)$ does not depend on y_n.

The differential equation of nth order has n parameters of initial values, which are in general different from each other. By fixing specific numbers for the initial values, one selects a specific function out of all the innumerable functions that satisfies the differential equation. The physical solution is thus obtained from the differential equations *and* the initial values.

The beauty and descriptive power of mathematics with its application in physics shines in a very specific manner for differential equations. One single, formally quite simple, relationship can include a multitude of solution possibilities, out of which the selection of a few parameters picks the specific solution. Understanding the relationships of differential equations, therefore, is much more important than the knowledge of a large number of formulas for limited problem areas.

The differential equation $y'' = -y$ describes all phenomena in connection with undamped, sine-shaped oscillations. A factor a ($y'' = -ay$) does not change anything fundamental, since scaling the time with $t = \sqrt{a}t$ transforms the differential equation back to $y'' = -y$. What is the visual meaning of this equation? The curvature y'' is equal to the negative function value. This means that, for a large positive function value, the curvature will finally reduce the function value, and for a negative value the absolute value finally decreases. However, this is exactly the hallmark of a periodic oscillation: the values do not go beyond a maximum or minimum value, but are always led from one to the other. For damped oscillations or those whose amplitude increases with time, one uses the more general law $y'' = ay' - by$. This states that for $a > 0$ the curvature increases with time, and thus the oscillation grows, while for $a < 0$ the curvature decreases, and thus the oscillation decays.

Other classes of phenomena can be described using different classes of differential equations. For example. the *Newton* equation of motion $m\frac{d^2\mathbf{r}}{dt^2} = \mathbf{F}(\mathbf{r})$ governs the huge class of all possible movements of a mass under the influence of a given force field $\mathbf{F}(\mathbf{r})$. This includes, for example, planetary movements. The resulting mechanical movements depend on the form of the force field and, particularly, on the initial values.

Thus differential equations can be viewed as condensed information that classifies a wide range of physical phenomena with some related characteristic. They also have a wide range of applicability as well as high aesthetic appeal, and provide order in the plethora of natural phenomena.

9.3 Solution methods for ordinary differential equations

If one wants to obtain from the differential equation a closed form for $y(t)$, then it needs to be solved *analytically*, as shown for the particularly easy case of the exponential function. If the function y does not appear on the right-hand side of the differential equation, i.e. if the differential equation reads $y' = f(t)$, then y is obtained via the normal *integration* method. We can immediately verify this for the traveled distance problem. For *constant* acceleration b we obtain for the traveled distance $s(t)$ the following:

$$s'' = b; \quad s' = v = \int_0^t b\,dt' = b\int_0^t dt' = bt + v_0;$$

$$s = \int_0^t (bt' + v_0)\,dt' = \frac{bt^2}{2} + v_0 t + s_0.$$

Here the two initial values are the initial position s_0 and the initial velocity v_0.

For the general case, the art of solving differential equations analytically fills entire books. The solution methods for those differential equations that are important in physics mostly follow simple patterns, for which there are standard methods. We refer here to a few of the books cited in the introduction. In general, all *ordinary* differential equations can be treated analytically. For this endeavor, an approach quite similar to the integration of non-standard functions is applied: one tries to guess a specific solution systematically and then tries to obtain a general solution with the variation or determination of parameters, which can be an exact, an approximate or a series solution of the differential equation.

If one does not need to obtain the solution as an analytical expression, but can be satisfied with calculating its numerical values as a function of the initial values and variables, and thus also to represent its general behavior graphically, then one can solve the differential equation numerically, irrespective of its complexity. All popular programs like *Mathematica* or *Java/EJS* provide a number of methods with different degrees of accuracy, which can be easily used. However, the algorithms use stay hidden in the background (black box), which is why we will be describing and visualizing the most important ones in the following paragraphs.

In practice, it is quite important to become familiar with the numerical methods of solution for *first order differential equations*, since all other ordinary differential equations can be reduced to them, if one allows several dependent variables. We visualize this in the following for equations of first order and also show in detail the application to differential equations of second order. All the following extensions work in a similar manner.

9.4   Numerical solution methods: initial value problem

If one also allows for nonlinear relationships, most of the differential equations that are important for physics are surprisingly simple. This may not really be a surprise, but nature in its deepest relationships is really simple! The causes and effects expressed via differential equations can thus be clearly and quickly derived and understood from physics.

In spite of this simplicity, the analytical solution of differential equations can become highly complex, especially if they are nonlinear. In all generality they can only be solved (integrated) in individual cases. Therefore, one makes simplifying assumptions about the form of the differential equations or first makes calculations for simply solvable special cases, and uses these to treat general cases that deviate only a little from these, using the so-called "perturbation theory".

Due to the ubiquity of the personal computer, many of these restrictions fall away. Suitable numerical programs can calculate the solutions of nonlinear and implicit differential equations as quickly and accurately as the solution of those equations, which can be easily solved in the classical analytical way.

Using a computer it is also possible to solve systems of many differential equations in reasonable time. An example of this would be where it is necessary to calculate the interaction between numerous bodies. In this case, an N-body problem leads in general to $6N$ differential equations and a corresponding number of initial conditions, since each body has six degrees of freedom, namely three for the position and three for the momentum. An extreme example is a simulation of the gravitational collapse in the early cosmos, which was conducted at the *Max Planck Institute for Astrophysics* in Munich.[18] In this calculation, the interaction between 10^{10} mass points was simulated and the calculation time of the supercomputer used was about one month! On the data carrier for the digital version of the essay edited by Martienssen and Röss, which was announced in the introduction to the present volume, there is a video of this simulation, which is described in a contribution by *Günter Hasinger*.

In the following, we sketch the general approach to the solution first for the example of the explicit first order differential equation, and then compare it with the familiar integration approach.

Given an initial value of the function and the relationship between derivative, function and variable we proceed as follows:

direct integration	*differential equation*
differential equation $y' = f(x)$	differential equation
	$$y' = f(x, y) \rightarrow y'_0 = f(0, y_0)$$
initial value $y_0 = C$	initial value y_0
solution $y = \int_0^x f(x)dx = g(x) + C$ with $f(x) = \frac{dg(x)}{dx}$.	required: solution for $y(x)$ with $y(0) = y_0$.

For the normal integration task, the derivative is a priori known in the whole interval as a function of the variable, while for the differential integration it can initially only be computed from the differential equation and the given initial value. For other values of x one does not yet know y, and therefore also y'. Thus one has to determine for the whole interval y and y' at the same time. To achieve this one has available, in addition to the differential equation, which is valid everywhere, only the initial value, together with the initial value of the derivative obtained from the differential equation.

The numerical methods correspond to a *careful step from the first to the second point*, from there to the next, and from there to the one after that, and so on. Thus, depending on the method, one arrives at a more or less suitable guess, as to where the next point could lie, given the initial values and the initial slope. For this point one uses the differential equation to calculate an estimate of the next point. In every step errors are created, and therefore it is quite astonishing what accuracy can be reached with

---

18 See Hasinger's Essay in *Physik im 21. Jahrhundert – Essays zum Stand der Physik*, Martienssen, W.; Röss, D. (editors), Springer, 2011, ISBN 978-3-642-05190-6.

advanced methods for rather simple algorithms. This is helped by the fact that many interesting tasks deal with periodic problems (orbits of planets, pendulums, periodic electric fields), where positive and negative errors from the two half periods often compensate for one another.

9.4.1 Explicit Euler method

For the simplest method, the classical *Euler* method, one assumes that the next value of y lies on the tangent that starts at the initial value y_0 and has a slope of $y'(0)$:

$$\text{initial value given as } y_0$$

$$\text{differential equation } y' = f(x, y) \rightarrow y_0' = f(x_0, y_0)$$

$$y_1 = y_0 + \Delta x \cdot y_0'$$

$$x_1 = x_0 + \Delta x \rightarrow y_1' = f(x_1, y_1)$$

$$y_2 = y_1 + \Delta x \cdot y_1'$$

$$\vdots$$

$$y_n = y_{n-1} + \Delta x \cdot y_{n-1}'$$

$$y_n' = f(x_n, y_n).$$

The method is called "explicit", since only data from the $(n - 1)$th point are used to calculate the nth point.

The Euler method is analogous to the integration of a known function y using the previously discussed method of trapezoidal steps. The additional complication with the analogous use for the initial value problem of a differential equation is that both the function as well as its derivative are unknown except for the initial point. The knowledge of the relationship between function and derivative is, however, sufficient to determine both of them approximately. However, one pays the price that the determination of y_1' at the first point is affected by the error committed when estimating y_1 itself from the initial values.

In Figure 9.5 the situation is clarified graphically for the example of the exponential function drawn in red. At the initial abscissa x_0, the initial value of the function y_0 is known. The differential equation yields the slope of the tangent drawn in blue. Its intersection with the interval boundary x_1 gives the next value according to the Euler method $\tilde{y}_1$ marked by a blue circle. In this example, this value is clearly smaller then the actual value y_1 of the exact curve, since the exponential function does not have a constant, but rather a constantly increasing, slope. The Euler method does not take into account changes in the derivative during the interval. Therefore one must make Δx as small as possible, to limit the error.

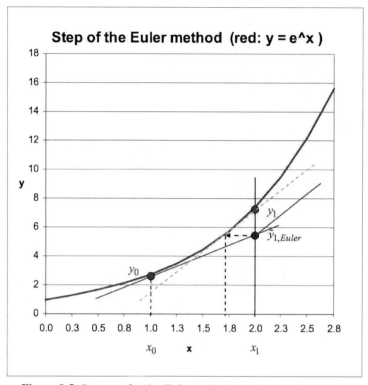

Figure 9.5. One step for the *Euler* method: see the text for details.

According to the construction of the algorithm, one does not use the slope of the curve at the initial point x_1 of the new interval, but a value that is obtained from abscissa and ordinate of the point via the differential equation, i.e. $y_1' = f(x_1, \tilde{y}_1)$.

We have chosen the exponential function as an example, since the ordinate x does not appear explicitly in the corresponding differential equation. Therefore, a simple graphical construction is possible for the second value of the derivative: it is equal to the slope of the dashed green tangent on the red curve at the ordinate of the second point $\tilde{y}_1$. With this slope we continue (blue) parallel to the dashed green line from the first calculated point to the next one.

In the general case the relationship would be less clear.

As known from the analogous integration method, the error in this simple method is quite substantial. It can be controlled to some extent at the price of larger computational effort by choosing the intervals Δx sufficiently small, and decreases linearly with the width of the integration intervals. *With growing resolution the method converges linearly to the correct solution.* For periodic functions, the errors partially compensate for each other in the half periods, since the deviation is negative for a concave graph, while it is positive for a convex one.

9.4.2 Heun method

The *Heun* method also calculates the corresponding next point in such a way that it lies ▢Heun on a straight line through the initial point x_0 (This also applies for the *Runge–Kutta* method, which is described next and has particular practical importance). In contrast to the *Euler* method, a more favorable angle is used. For the *Euler* method, this angle was simply determined as the result of the differential equation at the respective initial point of the new interval. For so-called *multi-stage* methods, of which the Heun method is one, this angle is determined as the mean value of several calculations. Thus the slope is obtained in more than one point using the differential equation.

As shown in Figure 9.6 the *Heun* method uses for one step of the method the differential equation both at the initial point and at the endpoint of the interval. As with the Euler method, it first calculates the so-called Euler point (the blue point in Figure 9.6) using the slope of the tangent at the start point. It then calculates the corresponding derivative at this point. In the figure this slope corresponds to the dashed blue tangent. Now the mean value of these two slopes (not of the angles, but rather of their tangents) is calculated, which is indicated by a dashed line with the corresponding slope in magenta. With this average slope, one now calculates in the forward direction from the initial point (solid green line, which has been shifted in parallel). Its intersection with the interval boundary at x_1 is the next point of the *Heun* method (green point). It is considerably closer to the "true" value than the result of the Euler approximation.

Expressed in formulas:

$$\text{forward} \qquad y'_0 = f(x_0, y_0)$$

$$y_{1,\text{Euler}} = y_0 + \Delta x \cdot y'_0 \qquad \text{Euler point as intermediate step}$$

$$y'_{1,\text{Euler}} = f(y_{1,\text{Euler}})$$

calculation of the mean value

$$\overline{y'_0} = \frac{y'_0 + y'_{1,\text{Euler}}}{2} \quad \rightarrow \quad y_1 = y_0 + \Delta x \cdot \overline{y'_0}; \ y'_1 = f(y_1).$$

In the form presented above, the Heun method is *implicit*, since the new point to be calculated appears on both sides of the equation. The equations therefore have to be solved with iterative methods.

The Heun method proceeds analogously to the integration of a known function with the help of the trapezoidal chord method. As shown when this method was discussed, the accuracy is considerably better then for the Euler method. The error of the Heun method thus decreases quadratically with the interval width, *the method converges quadratically*. It takes the change of the derivative within the interval into account in a linear approximation, thus it considers a *kink*.

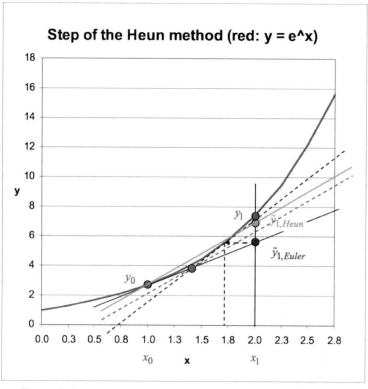

Figure 9.6. One step for the *Heun*-method: see the text for details.

9.4.3 Runge–Kutta method

The Euler and Heun methods have been described for historical systematic reasons, but even more so for didactic reasons. In their simple form they are no longer generally used, since the larger computational effort per interval for more advanced *multistage* methods is no longer an issue today, and therefore one can achieve much more accurate results for the same interval width.

The most popular route to the integration of differential equations is the *Runge–Kutta* method. In its four-step basic version it is analogous to the parabolic approximation for the integration of known functions, and takes into account the change of the slope within in the interval in a quadratic approximation; thus it uses a *parabolic curvature*. As for integration using parabolic approximation, *it converges with the fourth power of the interval width* $\propto \Delta x^4$.

For the parabolic method one uses, as described above, three points to fix the parabola, that approximate the true curvature in the interval: the initial point x_0, the midpoint of the interval $x_{1/2}$ and the endpoint x_1.

For the *integration*, the value of the derivative of the antiderivative to be determined is known across the whole interval, thus also at those three points. When solving the differential equation, the derivative is initially only known at the start-point of the first interval. The derivatives at the following points first need to be found. We now first compare for the start-point y_0 and the interval width $\Delta x \equiv \delta$ the structure of the formulas for calculating the next point y_1.

Integration using parabolic method　　　　Runge–Kutta method

$$y_0 = y(x_0)$$

$$y_1 = y_0 + \frac{\delta}{6}(y_0' + 4y_{1/2}' + y_1')$$

$$y_0 = y(x_0)$$

$$y_1 = y_0 + \frac{\delta}{6}(y_0' + 2y_{1/2}'^a + 2y_{1/2}'^b + y_1'^c).$$

One can recognize the formal similarity. However, for the *Runge–Kutta* method the listed derivatives are not the actual differential quotients of the desired solutions, but auxiliary variables which are obtained using the differential equation. In addition, we use instead of the derivative in the middle of the interval the mean value of two corresponding quantities with indices a, b.

Runge–Kutta method for an interval

interval width δ

initial variable x_0

initial ordinate y_0

$$y_0 = y(x_0) \qquad\qquad\qquad \rightarrow \qquad y_0' = f(x_0, y_0)$$

$$y_{1/2}^a = y_0 + \frac{\delta}{2}y_0' \qquad\qquad \rightarrow \qquad y_{1/2}'^a = f(x_0 + \frac{\delta}{2}, y_{1/2}^a)$$

$$y_{1/2}^b = y_0 + \frac{\delta}{2}y_{1/2}'^a \qquad\qquad \rightarrow \qquad y_{1/2}'^b = f(x_0 + \frac{\delta}{2}, y_{1/2}^b)$$

$$y_1^c = y_0 + \delta y_{1/2}'^b; \qquad\qquad \rightarrow \qquad y_1'^c = f(x_0 + \delta, y_{1/2}^b)$$

$$y_1 = y_0 + \frac{\delta}{6}(y_0' + 2y_{1/2}'^a + 2y_{1/2}'^b + y_1'^c) \qquad \rightarrow \qquad y_1' = f(x_0 + \delta, y_1).$$

One defines an auxiliary abscissa in the middle of the interval and calculates for it in a two step procedure two points a and b with their ordinates and derivatives. The first intermediate auxiliary point in the middle of the interval (index a) corresponds to the *Euler* point for half the interval width. Using the derivative at the Euler point one determines, beginning at the initial point, a second point in the middle of the interval (b). With the derivative at this point, one determines a third point at the end point of the interval with its associate derivative (c). After taking the average of the two derivatives at the midpoint, one has three points for the integration according to the parabolic method.

9.4.4 Further developments

The four-stage Runge–Kutta method described above converges so well that it is used for many applications.

One can improve the convergence of the method further by including additional points – similar to an approximation using polynomials of a higher order.

The speed of computation can be increased considerably by choosing the interval width not constant, but adapting it to the slope and curvature of the function to be integrated ("adaptive interval width"). This possibility is contained in popular five-stage Runge–Kutta programs and other numerical programs. Thus the interval can be selected automatically in such a way that a given error per interval is not exceeded.

The approximation rules used in the Runge–Kutta method have been well tested, but they are not the only feasible ones. One can work with other, more favorable, criteria, for specific classes of functions. In addition, there are quite a few approximation methods that have been derived in a very different way and that are also part of commercial programs and are discussed in the literature.

The computation speed of all methods depends on whether one works in higher-level languages or with languages that are closer to the operating system. Programs in *Java* or in *Mathematica* therefore run faster then algorithms written for example in *Visual Basic* for EXCEL. The speed of the following *Java* simulations is not limited by the computation speed, but is selected in such a way that one can easily follow the time development.

A program that one has written from scratch has, compared with using pre-built algorithms that run in the background, the didactic advantage that one can accurately follow the development and intervene in it.

9.5 Simulation of ordinary differential equations

9.5.1 Comparison of Euler, Heun and Runge–Kutta methods

The interactive image in Figure 9.7a leads to a simulation which shows the three methods in parallel for the example of the exponential function. The initial value for $x = 0$ is $y_0 = 1$ but can also be chosen differently. The number of intervals in the variable region can be chosen between 1 and 24.

There are four straight lines in the picture, which can be drawn and turned with the mouse. These allows the construction of the approximations to be easily visualized.

For the initially shown rough resolution one clearly recognizes the different convergence quality of the methods and the large superiority of the *Runge–Kutta* method – with the eye its error can no longer be noticed.

The description of the simulation contains further details and suggestions for experiments. It also contains a description of the complete codes, which are in each case a few lines that are repeated in a loop once for each point of computation. The

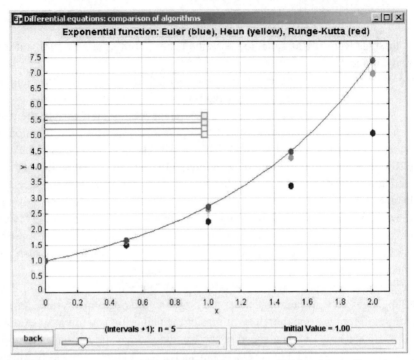

Figure 9.7a. Simulation. Comparison of the convergence of *Euler* (blue), *Heun* (green) and *Runge-Kutta* (red) methods for $y' = y$ (exponential function, blue line). The green lines can be pulled with the mouse to make the construction of the methods for one interval by hand possible. With a slider the number of intervals in the constant variable region can be changed (number of points n = number of intervals $+1$). With the second slider y_0 is adjusted (in the picture $y_0 = 1$).

calculation happens so quickly that one does not notice the time development in this example. For the commercial programs one can specify how many points must be calculated per minute in order to create, in the resulting graphs, the impression of a temporal sequence.

In practice nowadays one does not need to make the effort to write computational algorithms for the solution of ordinary differential equations, since they can simply be called up in all numerical programs by specifying a name. However, it is important that one understands how this "witchcraft" actually comes to be.

In Figure 9.7b the relative error of the three methods discussed in Figure 9.7a is shown, i.e. for example $(\tilde{y}_{\text{Euler}} - y_0 e^x)/y_0 e^x$. The ordinate region has been spread, such that the differences are more visible. For the small number of two to three points (one to two intervals) in the variable region even the small error in the Runge–Kutta method becomes noticeable in the plot. In order to also rate the error for a larger resolution, a number field of the simulation shows the relative deviation at the end of the last interval with high accuracy.

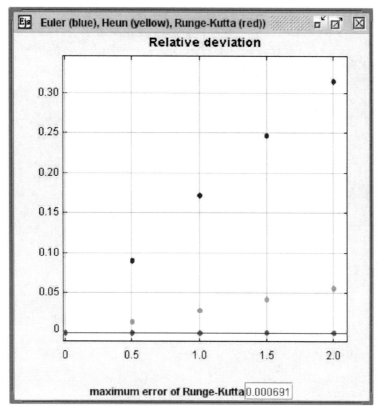

Figure 9.7b. Simulation. Comparison of the relative error for four intervals in the variable region (blue Euler, green Heun, red Runge–Kutta). The points show the relative deviations from the analytical value of the exponential function, which is shown with as a blue line. For the Runge–Kutta method the error at the end of the interval is shown in a number field. The scale on the y-axis depends on the accuracy achieved via the Euler method.

9.5.2   First order differential equations

We use here a *Runge–Kutta* procedure, which is integrated into *EJS*, to visualize explicit differential equations of first order. Implicit equations play a minor role in elementary physics. Their numerical solution can be achieved via iterations that are built into the computational algorithms.

In the graphs we use for the variable the symbol x and for the ordinate the symbol y.

The following interactive Figure 9.8a shows the graph of a transient process, which is defined by the differential equation that is shown in the text field y'. In this presentation the individual computation points are shown; one can switch to a line presentation using the option boxes.

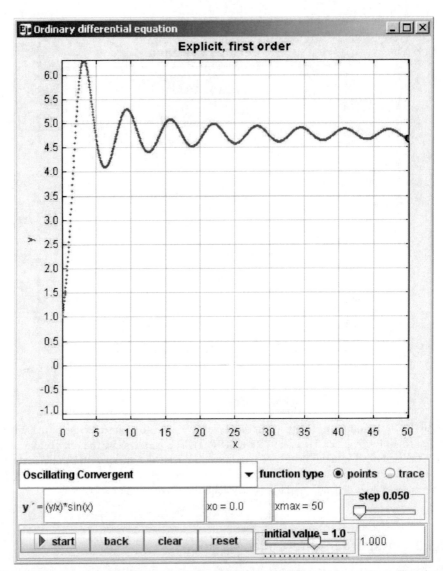

Figure 9.8a. Simulation. Animated solution of first order differential equations. The picture shows a convergent transient process. The range of the variable x, the initial value y_0 and the step width for the calculation can be chosen. One can select either a point presentation or a smoothing line presentation. With the option boxes one can select a new calculation or a superposition of calculations with different settings.

The differential equation shown in a text field can be edited or entered from scratch, so that you can investigate arbitrary explicit first order differential equations. The speed of the animation and the accuracy of the calculation that is related to it can be varied with the slider for the step width.

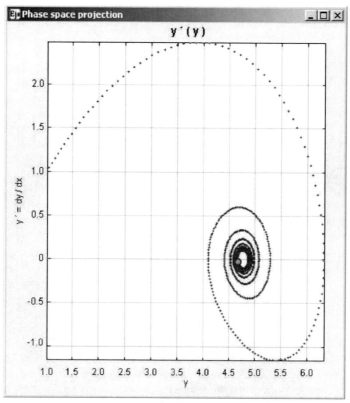

Figure 9.8b. Simulation. Phase-space y' versus y of first order differential equations for the example in Figure 9.8a. The green point shows the current computation point during the animation, which starts at $(1, 1)$.

The option box allows for selection among a number of elementary differential equations, which are preset with initial value, for example:

- exponential function $y' = y$;
- exponential decay $y' = -y$;
- transient processes;
- constant velocity $y' = C$ with C constant;
- constant acceleration $y' = Cx$ with C constant.

In the last two cases the solution of the differential equation is reduced to the normal integration process, since it does not contain y, and therefore these differential equations have as solutions the *anti-derivatives* of C and Cx.

The examples are classified according to the following characteristics:

- *divergent* (as the exponential function);
- *convergent* (as the exponential decay);

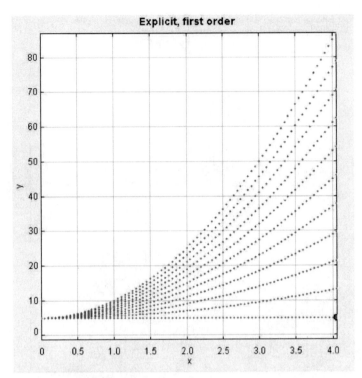

Figure 9.9. Family of solutions from the simulation in Figure 9.8a for the preset example $y' = ax$ (constant acceleration a) with parameter $a = 0, 1, \ldots, 10$ and $y_0 = 5$. Interpretation: $x =$ time (t), $y \equiv$ distance, $y_0 \equiv$ start position.

- *periodic*;
- *oscillating and divergent*;
- *oscillating and convergent*.

In Figure 9.8b a phase space projection y' versus y is shown. This shows the character of the differential equation and of its solution, *here oscillating convergent*, quite clearly. The green point designates the current endpoint of the calculation. In this example y converges against a finite value, while y' converges to zero.

The initial value y_0 and the initial abscissa x_0 can be chosen at will. The formulas are editable, such that you may enter arbitrary analytic functions and study them.

Multiple runs can be organized with the switches in order to compare the curves for different initial values, initial abscissae or differential equations. The passive picture in Figure 9.9 shows a simple example for constant acceleration a with the differential equation $y' = at$. Here the acceleration a is increased over 11 steps from 0 to 10. The initial value stays at $y_0 = 5$.

The description pages of the simulation contain further details and numerous suggestions for experiments.

During the initial work with the simulation one is often surprised by the totally unexpected results when entering a specific equation for y' or even only changing the value of a parameter in the equation. One is used to have a mental picture of dependencies of the form $y(x)$, but this is not the case for $y' = f(x, y)$, if one is not familiar with this.

Thorough experimentation with this simulation and the following examples for second order differential equations is therefore necessary to obtain a thorough understanding of the relationships that are described by the differential equations.

These examples demonstrate that a single differential equation defines a relationship, which contains an unlimited number of specific solutions. The *initial values* fix a particular solution from the family of possible solutions. The parameter that determines the specific solution does not have to be the initial value of the solution. One can also demand that the value of the function must be y_1 at a later time t_1. For the numerical solution one then solves the differential equation starting from y_1, first in the direction of increasing $t > t_1$ and then in the direction of decreasing $t < t_1$.

For a first order differential equation, the family of solutions has *one parameter*, for second order differential equations the family of solutions has *two parameters* (see the following subsection).

9.5.3 Second order differential equations

Numerous relationships in physics are described by second order differential equations. In addition to the acceleration (second derivative) they also enable you to take velocity dependent interactions into account (first derivative), which include friction processes. They also cover all undamped, purely periodic functions as special cases. The inclusion of damping makes realistic models of pendulums and oscillators possible.

The elementary functions described by first order differential equations are covered by analogous second order differential equations. We have already discussed how the differential equation of a similar structure will then contain *additional* functions. The solutions of second order differential equations constitute a *two parameter* family.

Only with *two* initial values, y_0 and y_0', for the start value x_0 of the variable, a specific solution is fixed. Thus a single initial value still allows an entire one-parameter family of solutions.

Among the explicit differential equations, a very important one is the simple equation whose solutions are the trigonometric functions:

$$y'' = -y \quad \text{or} \quad y'' + y = 0$$

$$
\begin{array}{lll}
y = \sin t & y = \cos t & y = e^{it} = \cos t + i \sin t \\
y' = \cos t & y' = -\sin t & y' = i e^{it} \\
y'' = -\sin t = -y & y'' = -\cos t = -y & y'' = i^2 e^{it} = -e^{it} = -y.
\end{array}
$$

It describes many oscillation processes.

As already discussed towards the end of Section 9.2, one reduces this differential equation for the numerical solution to a system of two coupled first order differential equations.

$$\text{general } y'' = f(y, y', x)$$

$$\text{1st definition: } y(x) \equiv y_1(x)$$

$$\text{2nd definition: } y' = y_1' \equiv y_2$$

$$\rightarrow y_2' = y_1'' = y'' = f(y_1, y_2, x).$$

The first equation defines the first of the new functions using the original function. The second equation defines the second new function as the derivative of the first one. The original differential equation connects y_2' with y_1, y_2 and x. From the solution for y_1, y_2 one recovers y and y'.

Thus the two coupled first order differential equations

(a) $y_1' = y_2$

(b) $y_2' = f(y_1, y_2, x)$ for the two functions y_1, y_2

are equivalent to the single differential equations of second order for $y(x)$

$$y'' = f(y, y', x).$$

special case: $y'' = -y$ via $y \equiv y_1$ and $y' = y_1' = y_2$ becomes

the system of two differential equations for y_1 and y_2

(a) $y_1' = y_2$;   (b) $y_2' = -y_1$.

The steps by which the two equations are solved for subsequent points have to be nested in a suitable way. Using equation (a) one first calculates an approximation for the derivative, which is then substituted in equation (b) instead of the formally required derivative. In practice this algorithm is contained in all popular numerical programs. We again use for our examples an *EJS* simulation, for which we only add equation (a) in an additional line. (As designation for the first derivative we use in the formula field "yStrich" ($\approx$ yprime) since *Java* cannot understand "y'".)

For a differential equation of higher order this method would have to be repeated for every further order and chained in an equivalent manner. Differential equations of higher order do not, however, play a major role in physics.

The following interactive picture in Figure 9.10a leads to a simulation for *second order differential equations*. It shows an exponentially damped periodical oscillation. In the differential equation $y'' = -y' - 0.2y$ shown in the text box the first term $-y$ is responsible for generating a periodic function and the second term $-y'$ for the

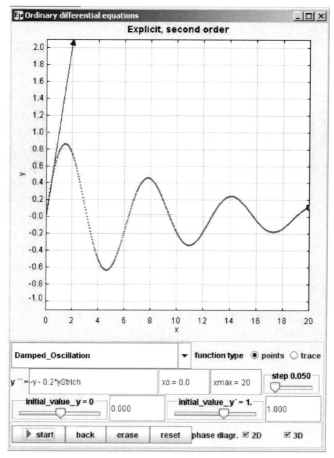

Figure 9.10a. Simulation. Animated simulation of the solution of second order differential equations. Example: damped oscillation with the initial values $y_0 = 0$ and $y'_0 = 1$. The arrow shows the two initial values for value and derivative. Variable range, initial values, step width and presentation type can be set. In addition, the phase space diagrams can be shown in a 2D or a 3D presentation.

exponential decay, as familiar from the first order differential equations. The factor 0.2 determines the speed of decay.

The control of this simulation is quite similar to the case of first order differential equations; only control elements for the second initial value y' are added. In the selection box differential equations and initial values for the following functions are preset:

- cosine
- sine
- exponential function

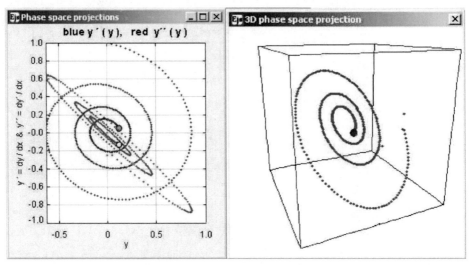

Figure 9.10b. Animated phase spaces for the differential equations of Figure 9.10a. The left windows plots $y'(y)$ in blue and $y''(y)$ in red. The respective end points are highlighted and marked in color. In the right windows y, y' and y'' are mapped to the three space axes. The red curve thus represents the total differential equation $y'' = f(y, y')$. The 3D projection is rotated in the animation.

- exponential damping
- hyperbolic sine
- hyperbolic cosine
- delayed oscillation
- accelerated oscillation
- damped oscillation
- growing oscillation.

All parameters can be changed. In the text field the differential equation can be changed or a totally new one can be entered, such that you may investigate arbitrary second order differential equations using this simulation.

With two switches a 2D presentation $y'(y)$ and $y''(y)$ and/or a rotating 3D presentation $y''(y, y')$ of the phase spaces can be chosen. This window of the simulation is shown in Figure 9.10b

The two-dimensional phase diagram now shows two curves $y'(y)$ in red and $y''(y)$ in blue. In this example one recognizes the damped transient process as a double exponential spiral.

The three-dimensional phase diagram shows $y'' = f(y, y')$ as a plane spiral in phase space. Its rotation during the animation increases the spatial impression.

The description pages of the simulation file contain further details and suggestions for experiments.

9.5.4 Differential equations for oscillators and the gravity pendulum

The second order differential equations discussed in Section 9.5.3 describe, among other systems, all possible kinds of oscillator, including also the classical mathematical gravity pendulum (called *mathematical* because it treats the pendulum as a mass point on a mass-less stiff rod in abstraction from its real construction). For these cases, the differential equations and initial conditions of the following simulation are pre-formulated but, in other respects, the simulation is very similar to the previous one.

The interactive image in Figure 9.11a shows the example of a damped oscillator, which initially oscillates in its eigen frequency until $x = 30$, when an external force at double the frequency is added to the system. One sees the transition from the free oscillation to the forced oscillation at double the frequency including interferences. The free oscillation finally decays away totally. The driven oscillation remains with double the frequency and a constant amplitude.

The corresponding phase space curve in Figure 9.11b is quite confusing as a static picture. If, however, one observes the dynamic flow, one recognizes the different transitions quite easily.

When cleared of factors that scale the graphics or are needed for the formula to be recognized (*yStrich* instead of y'), the differential equation reads: $y'' = -y - y' + \sin 2x \, \text{step}(x - 30)$.

The term $-y$ produces a periodic oscillation with period 2π, the term $-y'$ an exponential damping and the term $\sin 2x$ a driving force with constant amplitude and the period π. The very useful *step* function switches at the given point in time $x = 30$ from 0 to 1. The damped oscillation of the free pendulum simply continues, while the periodic driving force is added at this point.

In the phase space diagram shown in Figure 9.11b, one also recognizes the transition between the two kinds of oscillations, from the initially free and damped oscillation (initial plane spiral) to the forced oscillation. After a sufficiently long time, the free oscillation has been damped away and the oscillator moves periodically with constant amplitude a the frequency of the driving force.

The simulation contains the following pre-defined oscillators:

- free oscillator with adjustable eigen frequency;
- dissonant driving force with adjustable frequency;
- resonant driving force;
- dissonant driving force with damping;
- resonant driving force with damping.

In addition, for the gravitation pendulum as second pendulum (full period of 2 s at small amplitude), the following situations are preset:

- deflection of a few degrees;

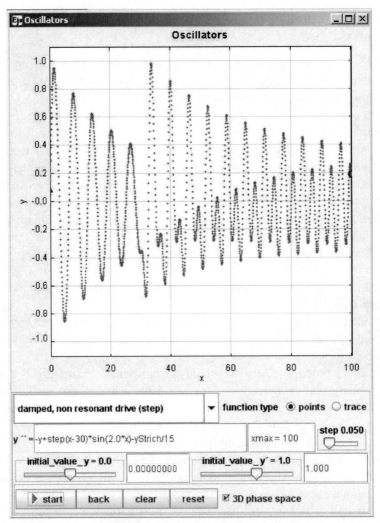

Figure 9.11a. Simulation. The figure shows the solution of an oscillation equation with damping, which is driven by an external force at double the eigen frequency and supplied with energy starting at $x = 30$. The differential equation and all parameters can be changed.

- deflection nearly up to the rollover, i.e. angular deflection from the rest point of nearly π;
- shortly after the rollover, i.e. residual velocity at the turning point.

The plots of phase space curves for the gravity pendulum in the passive Figure 9.11c show in the left window the situation for a deflection of 5.7 degrees, for which the oscillation is still practically sinusoidal (red curve $y'' \approx -y$) and in the right window

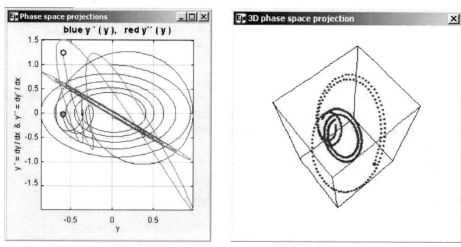

Figure 9.11b. Phase space plots for the oscillation equation of Figure 9.11a. On the left projections y' versus y in blue and y'' versus y in red, on the right y versus y' and y''. The picture shows the state shortly after adding the external driving force, on the left as lines, on the right as sequence of calculated points.

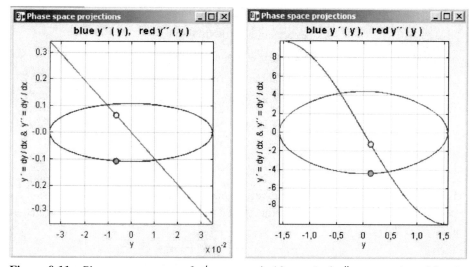

Figure 9.11c. Phase space curves of y' versus y in blue and of y'' versus y in red for the pendulum example of the simulation in Figure 9.11a: shown on the left for small and on the right for large deflections. Please note the different scales on the axes, especially the ordinate scale. The red line in the left window shows the negative linear relationship between acceleration and angle of deflection. In the right window one recognizes the large nonlinearity for large deflections. Therefore only pendulums with deflections of a few degrees can be used for accurate clocks.

for a deflection of 90 degrees, for which the oscillation deviates quite clearly from it. Thus the blue curve is therefore for the small deflection a circle which is traversed with constant angular velocity (please note the different scales on the axes!). For the large deflection on recognizes in the animated simulation the extended time spent in the vicinity of the turning points.

The formulas and initial values can again be changed. In the vicinity of the unstable equilibrium (deflection of π) the solutions become extremely sensitive to the initial values, but also to the accuracy of the computation, which can be adjusted with the step width slider.

The description pages contain again exact details and hints for experiments.

9.5.5 Character of ordinary linear differential equations

From the experimental analysis of different explicit linear second order differential equations we can draw a few general conclusions:

the following term in the differential equation means respectively:

$y'' = -y$	$\rightarrow$	periodic function with period 2π
$y'' = -a^2 y$	$\rightarrow$	periodic function with period $2\pi a$
$y'' = -y'$	$\rightarrow$	exponential decay with x
$y'' = y$	$\rightarrow$	exponential growth with x
$y'' = y'$	$\rightarrow$	exponential change with x
$y'' = $ const.	$\rightarrow$	constant acceleration
$y'' = 0$	$\rightarrow$	constant velocity (0 acceleration)
$y'' = f(x)$	$\rightarrow$	x-dependent driving force, characterized by $f(x)$
$y'' = -yf(x)$	$\rightarrow$	periodic oscillation, moderated by $f(x)$
$y'' = -y'g(x)$	$\rightarrow$	exponential decay, moderated by $g(x)$.

The points to which convergent or divergent solutions move in the phase diagram are referred to as *point attractors* and the closed target curves of periodic solutions are called *periodic attractors*.

9.5.6 Chaotic solutions of coupled differential equations

A new phenomenon appears if three or more first order differential equations are cou- pled and contain terms that are nonlinear in the variables. For certain parameter regions or regions of the initial values, or even for all initial values, their solutions show chaotic behavior. This is especially attractive for oscillating systems that are characterized by second order differential equations with the fundamental dependence $y'' = -y \pm \cdots$.

Driven double pendulum

As a first example we want to investigate in Figure 9.12a the simulation of a double pendulum, for which a second mathematical pendulum is fixed to the end of a primary mathematical pendulum (*mathematical* means here: the total mass is concentrated at the end of a stiff mass- and weightless pendulum rod).

The primary pendulum can be driven by a periodic force. The secondary pendulum is driven by the primary pendulum. Both are subject to gravity.

Each pendulum is described by a second order ordinary differential equation, which corresponds to four first order differential equations, and the differential equations are coupled; thus they also contain variables of the corresponding other pendulum. It is now essential that these differential equations are coupled by trigonometric functions and quadratic terms:

$$y_1'' = f_1(y_1, \sin y_2, y_2, \sin(y_2 - y_1), y_1'^2, y_2'^2)$$
$$y_2'' = f_2(y_1, \sin y_1, y_2, \sin(y_2 - y_1), y_1'^2, y_2'^2).$$

The exact formulas are discussed in the description pages of the simulation.

The ratio of the pendulum lengths and the pendulum masses can be adjusted as well as the speed of the animation.

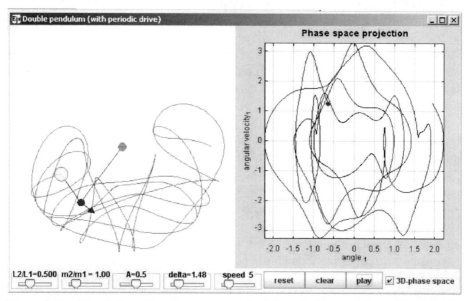

Figure 9.12a. Simulation. Chaotic movement of a driven double pendulum with adjustable length and masses (red curve). In the left window the double pendulum is shown (pivot point in green, mass point of the primary pendulum in blue, mass point of the secondary pendulum in yellow, vector of the external driving force as a blue arrow). In the right window the phase space projection angular velocity $d\phi/dt$ versus angle of deflection ϕ is shown.

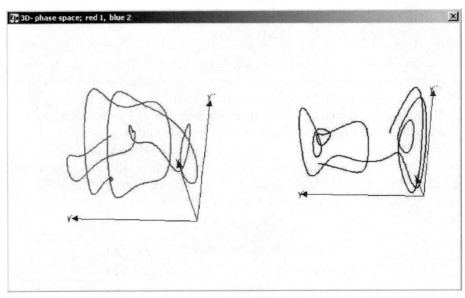

Figure 9.12b. 3D phase space diagrams for the double pendulum, on the left red for the end of the primary pendulum and on the right for the end of the double pendulum.

The indirect external driving force modulates the angular velocity of the primary pendulum with a sine function of adjustable frequency. The blue arrow shows direction and absolute value of the external driving force.

The red curve shows the orbit of the secondary pendulum, that is, of the mass point at the end of the mass-less pendulum rod of length l_2. It is possible to superimpose orbits for different initial values and thus to study at the same time the influence of small changes in the initial conditions on the long-term behavior.

In the right coordinate system of Figure 9.12a a plane phase space projection for the orbit of the primary pendulum is plotted. In addition a rotating presentation of the three-dimensional phase space y'' versus y' versus y can be switched on (see Figure 9.12b).

There is obviously no *periodic attractor*. One refers to a *strange attractor* if the phase space orbits of the process described are limited to a certain region of the phase space, and do not become periodic, but show a fractal character and therefore cannot be described in an analytic closed form.

Together with adjusting the ratios of pendulum length and mass, one obtains a rich spectrum of oscillation processes that happen chaotically but strictly deterministically.

The *Reset* button resets the simulation exactly (within the accuracy of the PC) to the same initial conditions. You may convince yourself that the time development, which looks so confused, is indeed repeated, and thus happens deterministically and not controlled by chance (this observation is achieved easily by calling the simulation twice and letting it run twice).

You may, however, also adjust the initial conditions for the position manually by pulling the yellow point; you will not achieve an exact reproduction, and two simulations running in parallel will soon run apart from each other. Thus the chaotic–deterministic character is connected to an extreme dependence on the initial conditions.

The simulation description contains numerous suggestions for experiments.

For the double pendulum with its many nonlinear connections, it will be barely possible to find a setting that leads to a periodic solution.

In other cases there are regions of chaotic behavior next to regions of periodic behavior.

Reflection of a ball between sloping walls

For the second example, shown in Figure 9.13, it is obvious that there must also be periodic solutions.

In this *ball in wedge* simulation, a ball is reflected back and forth between two infinitely extended planes. For an initial orbit that starts symmetrically to the axis of symmetry and ends orthogonal to one of the surfaces, the orbit is already closed after hitting both surfaces once. It can be suspected that there are further periodic orbits with many reflections. In general, however, the orbits are chaotic. The pitch of the surfaces and the position and initial velocity of the ball can be adjusted by pulling

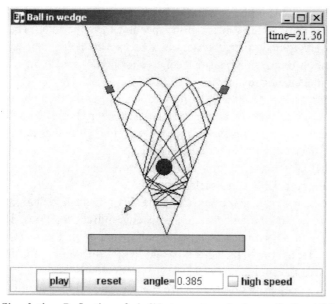

Figure 9.13. Simulation. Reflection of a ball between two sloping walls. In addition to chaotic orbits there also are periodic solutions, for which the ball jumps regularly back and forth on periodical orbits between the walls.

with the mouse. The nonlinearity of the connections lies here in the trigonometric functions used in the three coupled first order differential equations.

The example also demonstrates the use of the *Poincaré section* for the visualization of chaotic or periodic orbits and its use for the determination of periodic initial conditions. It shows the intersection point of the orbit in the symmetry plane. Periodic orbits lead to a finite number of intersection points with regular patterns.

Many-body problem of gravitation

Chaotic behavior is not only an interesting theoretical problem, but is of large practical importance, since many phenomena in physics and engineering are described by more than two coupled nonlinear differential equations. This includes, for example, the gravitational processes in three dimensions. In this case the differential equations are nonlinear and of the following type for each Cartesian coordinate:

$$y'' = -gM\frac{y}{r^3} = -gM\frac{y}{[x^2 + y^2 + z^2]^{3/2}}.$$

Because of the basic type $y'' = -y$, one expects that periodic oscillations (orbits) should be possible for certain initial conditions. This is indeed the case for two bodies (in addition there are the cases of a collision for finite size of the bodies and the "scattering" for the case of a body that passes by). For three and more bodies there exist, except for very specific initial conditions, no long-term periodic orbits, but only more or less chaotic orbits, which can sometimes become quasi-periodic. The apparent regularity of the many-body planet system is a deception. This is due to the relatively short observation time, which is small relative to the time scale, in which the orbits will develop in a chaotic manner.

The situation becomes a bit simpler if one assumes, for the theoretical computation, that all bodies move in one plane, since then the number of coupled differential equations becomes smaller. If one assumes in addition that all bodies have the same size and the same mass m, one can for certain very specific initial conditions (symmetric configurations) also create periodic orbits for more than two bodies. The following simulation in Figure 9.14 shows such special cases.

Different scenarios can be selected using the slider on the left. One can pull individual bodies with the mouse and change the specific initial conditions. This leads very soon to a decay of the symmetric configuration. In addition, it turns out that even under such artificial assumptions there exists no long-term stability for more than three bodies, provided the simulation proceeds for a sufficiently long time. The following development can be easily observed when one zooms into the picture with the slider on the right.

How could any relatively stable and bound systems develop in the cosmos under these circumstances? One has to consider this as the result of a long-term evolution, with a multitude of collisions and disintegrations that provide for "friction", from

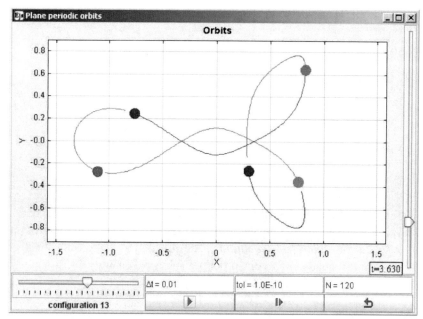

Figure 9.14. Simulation. Stable and non-stable solutions of the many-body problem of gravitation for movement in a plane: in the figure an example with five equal masses is shown. With the left slider different initial patterns can be chosen. The right slider allows you to zoom in or out of the image. The initial position of the bodies can be pulled with the mouse.

which those remnants are bound for a limited time to quasi-periodic orbits, which satisfy suitable initial conditions. Most remnants, however, vanished into the distance, until they interacted again with other systems under the exchange of energy. On the other hand, new candidates entered the original region from other regions leading to a change of initial conditions.

At this point you should again study the essay by Siegfried Grossmann,[19] who studies the question of chaotic systems very thoroughly. In the contribution by Guenther Hasinger you can see in detail how chaos and collisions can at least lead to order and structure in the cosmos for limited time periods. The simulations by Eugene Butikov simulate a wealth of many-body problems with partially periodic and partially chaotic behavior.

---

19 *Physik im 21. Jahrhundert: Essays zum Stand der Physik* edited by Werner Martienssen and Dieter Röss, Springer Berlin 2010.

10 Partial differential equations

10.1 Some important partial differential equations in physics

Physical events Φ generally take place in the three space dimensions x, y, z and the time t: $\Phi = \Phi(x, y, z, t)$. The spatial and time development are coupled to each other. The differential equations describing the phenomena then contain partial derivatives with respect to the space coordinates and with respect to time, and therefore are referred to as *partial differential equations*. The functional relationship for a general partial differential equation of second order for a physical quantity $\Phi(x, y, z, y)$ reads:

$$F\left(\frac{\partial^2 \Phi}{\partial \alpha^2}, \frac{\partial \Phi^2}{\partial \alpha \partial \beta}, \frac{\partial \Phi}{\partial \alpha}, \Phi, x, y, z, t\right) = 0$$

with $\alpha = x$ or y or z or t and $\beta = x$ or y or z or t, $\beta \neq \alpha$.

To keep the presentation readable, we have not shown all terms in the brackets, but only one of each type. This type is characterized by one of the variables or, in the case of mixed derivatives, by two of them. Thus, in addition to first and second partial derivatives, all mixed second derivatives can also appear.

Fortunately, the partial differential equations important in physics and engineering are much simpler than this general form, as the following examples will show. They are, however, still rather complicated and only allow an analytical solution and simple interpretation in very elementary cases. In the following we only cite a few important partial differential equations in physics and want to make you aware of the crucial differences between the boundary value problem/initial value problem for ordinary and for partial differential equations. For further information we refer to the specialist literature.

The simulation examples show specific solutions of the corresponding *one-dimensional*:

- diffusion equation for point-like initial impulse (delta impulse);
- Schrödinger equation for a point mass and for different oscillators;
- wave equation for a vibrating string.

a) wave equation

$\Phi(x, y, z, t)$ describes the deviation of the physical quantity at time t, for example of the field strength, the pressure and so on.

The differential equation reads $\dfrac{\partial^2 \Phi}{d(ct)^2} = \dfrac{\partial^2 \Phi}{dx^2} + \dfrac{\partial^2 \Phi}{dy^2} + \dfrac{\partial^2 \Phi}{dz^2}$,

and in one dimension: $\dfrac{\partial^2 \Phi}{dt^2} = c^2 \dfrac{\partial^2 \Phi}{dx^2}$.

The general, easy to check solution is then $\Phi(x,t) = f(x + ct) + g(x - ct)$.

The very general one-dimensional solution of the wave equation contains two arbitrary functions $f(x)$ and $g(x)$, which are propagated along the x-axis with velocity c in a negative/positive direction without changing their form. This example already demonstrates an important difference to solutions of second order ordinary differential equations: while there a solution was determined via two initial *values* y_0, y_0', it is now fixed via two initial functions $g(x,0)$ and $f(x,0)$. The second order ordinary differential equation has as the solution a family of functions with two arbitrary *number parameters*. The solution of this partial differential equation is described via a family of functions with two *initial functions*. For the one-dimensional case those are defined along the x-axis, for example a wave packet, in the simplest case a *sine wave* of undetermined position or a *Gaussian impulse*.

In the three-dimensional case, initial functions can be defined on a boundary, a surrounding surface, or in a volume.

b) one-dimensional heat conduction equation

heat equation

The field $\Phi(x,t)$ is here the temperature that depends on space and time coordinates.

The differential equation reads $\dfrac{\partial \Phi}{\partial t} = a \dfrac{\partial^2 \Phi}{\partial x^2}$.

For its analytical solution for a delta-pulse as initial function one obtains:

$$K(x,t) = \dfrac{1}{\sqrt{4\pi a t}} e^{-\frac{x^2}{4at}}.$$

The heat conduction equation, also called the diffusion equation, describes equilibration processes in time (here along a line, the x-axis). The special solution $K(x,t)$ in the example starts with the *delta* impulse as initial function. This means that the total heat is first concentrated in the point $x = 0$. This amount is then spread over time as a Gaussian distribution, while the integral (the amount of heat) stays the same. Thus the temperature maximum at $x = 0$ decreases accordingly.

c) Schrödinger equation

Schrö-dinger

The probability amplitude or wave function is $\psi(x, y, z, t)$;

and the potential is $V(x, y, z.t)$;

$$\dfrac{ih}{2\pi}\dfrac{\partial}{dt} = -\left(\dfrac{h}{2\pi}\right)^2 \dfrac{1}{2m}\left(\dfrac{\partial^2}{dx^2} + \dfrac{\partial^2}{dy^2} + \dfrac{\partial^2}{dz^2}\right) + V\psi.$$

The form of the Schrödinger equation given above is valid in the non-relativistic case for a particle of mass m in a potential V. It describes the relationship between time and space development of its complex wave function   .

d) Maxwell equations for the electromagnetic fields $\mathbf{E}, \mathbf{D}, \mathbf{B}, \mathbf{H}$

$$\text{1)} \qquad \text{div } \mathbf{D} = \rho \qquad\qquad \nabla \cdot \mathbf{D} = \rho,$$

$$\text{2)} \qquad \text{div } \mathbf{B} = 0 \qquad\qquad \nabla \cdot \mathbf{B} = 0,$$

$$\text{3)} \quad \mathbf{curl\,E} + \frac{\partial \mathbf{B}}{dt} = 0 \qquad \nabla \times \mathbf{E} + \frac{\partial \mathbf{B}}{dt} = 0,$$

$$\text{4)} \qquad \mathbf{curl\,H} = \mathbf{j} + \frac{\partial \mathbf{D}}{dt} \qquad \nabla \times \mathbf{H} = \mathbf{j} + \frac{\partial \mathbf{D}}{dt}.$$

The Maxwell equations, which are very important in practice, describe the interaction between the magnetic and electric fields (2 and 3) and their connection with the charge density ρ and current density $\mathbf{j}$. The first equation means that charges are the sources of the electric fields, from which field lines emanate and where they end. The second equation means that magnetic sources (monopoles) do not exist and therefore magnetic field lines are always closed.

On the left, the traditional notation, and on the right the formally quite uniform notation with the nabla operator, are given.

The electrical flux density $\mathbf{D}$ is connected to the electrical field strength $\mathbf{E}$ via the material properties *electrical permeability of the vacuum* ε_0 and *electric polarization* $\mathbf{P}$: $\mathbf{D} = \varepsilon_0 \mathbf{E} + \mathbf{P}$

The magnetic flux density $\mathbf{B}$ is connected to the magnetic field strength $\mathbf{H}$ via the material properties *magnetic permeability of the vacuum* μ_0 and *magnetic polarization* $\mathbf{J}$ (written in capitals, as opposed to the current density $\mathbf{j}$): $\mathbf{B} = \mu_0 \mathbf{H} + \mathbf{J}$

Since $\mathbf{D}, \mathbf{B}, \mathbf{E}$ and $\mathbf{H}$ are vectors, we have to deal with a system of coupled partial differential equations for all field components, which therefore has a wealth of solutions. Therefore the mathematical solution can become very complex.

Numerical solution methods are therefore even more important for partial differential equations than for ordinary differential equations. While one starts for ordinary differential equations from one or more initial values and iteratively proceeds from point to point for the independent variable, one has to cover the whole space of variables with a grid of computation points. For a two-dimensional problem one then deals with a plane grid and for a three-dimensional one with a three-dimensional space grid. One starts from one point of the initial function, calculates the neighboring points using suitable procedures, which together constitute the initial values for the next step, always while taking into account the connections provided by the differential equations. In technical applications and engineering one refers in this connection to the method of *finite elements*.

For visualization one simplifies the conditions radically. Already in Section 8.5.8 we had simulated the movement of an electron in a three-dimensional homogeneous electromagnetic field, which is stationary, i.e. is constant as a function of time.

10.2 Simulation of the diffusion equation

The following simulation in Figure 10.1 of one-dimensional equilibration or diffusion processes shows, for example, the time and space dependence of the temperature after heating a homogeneous thermally insulated thin wire at a point with a short pulse.

According to the above mentioned special solution, an approximated delta function at the origin is used as the initial function, which spreads in Gaussian shape under conservation of the area under the curve (the amount of heat). The arrows indicate the $1/e$-width, the number field the respective point in time. The diffusion constant a can be adjusted with the slider over a wide range of values.

The description pages contain further hints.

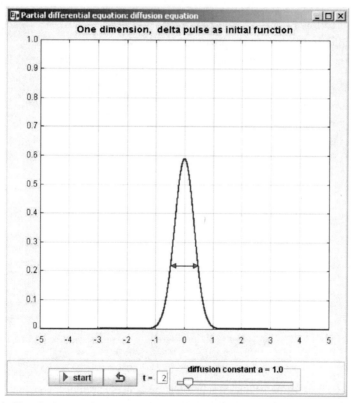

Figure 10.1. Simulation. Animated solution of the diffusion equation with the delta impulse for $t = 0$ at $x = 0$. The picture shows the state at $t = 2$. The arrow indicates the width, where the function has decayed to $1/e$ of the maximum.

10.3   Simulation of the Schrödinger equation

The interactive Figure 10.2a show the solution to the one-dimensional Schrödinger equation for a particle in an infinitely deep rectangular potential well, whose width can be adjusted with a slider. The square of the absolute value of the complex wave function $|\psi(x)|^2$ gives the probability density for the particle at position x. It is normalized to 1, which means that the particle can be found inside the box with certainty, irrespective of the spatial distribution.

The two curves in Figure 10.2a show the real component of the wave function (probability amplitude) $\psi(x)$ in red and the imaginary component in blue.

In Figure 10.2b, a second presentation mode that is popular in quantum mechanics is used, for which the absolute value of the wave function $|\ \ |$ (square root of the probability density) is shown as the envelope. Inside, the phase angle $\alpha = \arctan(\frac{\text{Im}}{\text{Re}})$ is indicated by color shading.

The phase angle α is indicated by the following colors:

- blue $\alpha = 0$ or 2π (positive real);
- golden yellow $\alpha = \pi$ (real negative);
- rose coloured $\alpha = \pi/2$ (positive imaginary);
- green $\alpha = 3\pi/2$ (negative imaginary).

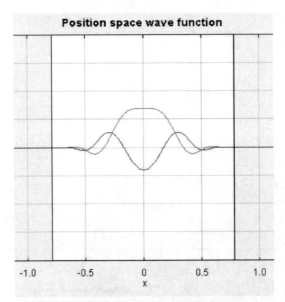

Figure 10.2a. Simulation. Animated solution $\psi(x)$ of the Schrödinger equation for the development of an initial distribution (symmetric Gaussian) in a box. The real component is in red and the imaginary component in blue. The probability density consists of the sum of squares of these two parts: $\psi(x)^2 = (\text{Re }\psi(x))^2 + (\text{Im }\psi(x))^2$.

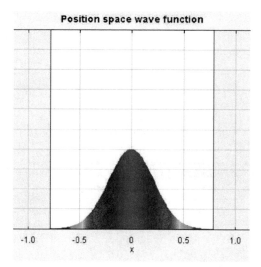

Position space wave function

Figure 10.2b. Simulation. The function $|\psi(x)|$ for Figure 10.2a; The color shading indicates the ratio of imaginary to real components. $|\psi(x)| = \sqrt{\text{probability density}}$. In the blue inner region the real component dominates, while the imaginary component dominates in the red regions.

This simulation allows you to choose among many examples of potential wells, in which quantum particles can move. It was developed by the pioneer of the *OSP program*, Wolfgang Christian, and slightly simplified by us. The description pages in Figure 11.2 contain detailed hints about theory and usage.

10.4 Simulation of the wave equation for a vibrating string

At the end of this chapter we consider in Figure 10.3b the simulation of a vibrating string as a solution of the wave equation. Figure 10.3a shows three snapshots as examples from the simulation in Figure 10.3b. The left-hand panel shows the "start impulse", a Gaussian concentrated in the middle of the string with maximum 1. The middle panel follows the situation shortly after the start: two Gaussian impulses of height $1/2$ run into opposite directions. They are finally reflected at the ends of the string and interfere with each other in the right-hand panel, which results in the reconstruction of the original form and amplitude, but with a negative sign after the first reflection.

Gaussian impulses and symmetric wave functions propagate on the string unchanged provided that no damping is taken into account in the wave equation.

The interactive Figure 10.3b show the situation a short time after the start of the triangular impulse, which was originally concentrated at the end of the string. After

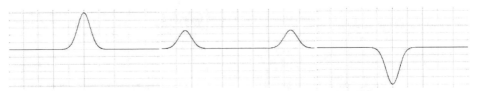

Figure 10.3a. Propagation of an excitation on a string that is fixed at both ends; Gaussian as initial impulse for $t = 0$ on the left; two pulses run in opposite direction in the middle; on the right the reconstruction with negative amplitude after reflection at the ends.

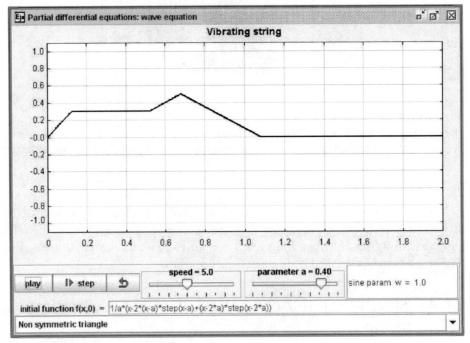

Figure 10.3b. Simulation. Time development of an original short triangular pulse at the end of the string. The picture shows the state after the first reflection at the end of the string.

some time one observes deviations, which are due to the discontinuity of the first derivative at the beginning and end of the impulse. This demonstrates limits of the numerical computation.

A selection menu contains the following start functions for the initial deflection of the string:

- Gaussian impulse of adjustable within the middle of the string;
- Gaussian impulse not in the middle of the string;
- symmetric triangle in the middle of the string;
- triangle at the end of the string;

- sawtooth;
- sawtooth of adjustable width;
- sine wave.

There is a parameter a in most of the functions, which can be changed. The formulas for the initial deflections themselves are editable, so that many more start situations can be simulated.

The description pages contain further details and suggestions for experiments. This animation is aesthetically quite pleasing, because it gives the music lover hints about tone qualities, which are possible as a result of very different overtone mixtures. More details are given in the description pages.

This simulation was originally developed by Francisco Esquembre, the pioneer of the *EJS program* and extended by us.

11 Appendix
Collection of physics simulations

11.1 Simulations via OSP/EJS programs

The interactive simulation of complex mathematical or physical objects is an attractive programming task, which provides a deeper understanding of the problems treated. It is a wonderful tool for the illustration of objects that are abstract or difficult to imagine, or which can only be calculated with difficulty by hand. It appeals to the playfulness of the user and leads to intensive preoccupation with the subject. Therefore, simulation can be viewed as an effective didactic tool. Great efforts have been directed to this end and the German ministry of research and education has, to some extent, supported these efforts financially.

Commonly used high-level standard programs, such as **Microsoft Excel** with **Visual Basic for Applications** (**VBA**), were candidates to use for these simulations. One finds many interesting examples of this approach on the Internet. A fundamental advantage of such technically relatively simple programming is that the user has open access to the code via the standard program. In addition, samples that are taken by third parties are therefore reasonably transparent, and the user can thus develop or modify the code, if the manufacturer has not built in artificial barriers. A major disadvantage is that files created in this way are platform dependent, and thus run only if the same operating system and application program (which are subject to licensing) are used. It even turns out that successive versions of standard application programs are not fully compatible. For example, a file developed with *Excel 2005* and *VBA* under *Windows XP* can be incorrectly formatted if run on another computer with *Windows Vista*. A technical disadvantage is that the computing speed of a program developed in a high-level tool such as *Excel* is much lower than that of a low-level program running close to operating system level for the same task.

Therefore, efforts were made early on to use platform-independent and operating system-level programming languages, and the *Java* royalty-free language is considered to be particularly suitable. However, simulation with common object-oriented *Java* requires programming experience of considerable depth. It would have appeared justified to narrow down *Java* in line with the limited issues involved. Unfortunately, there has been no systematic effort in Germany for defining a standard for mathematical and physical simulations. Rather, the results of different schools have originated more or less independently of one another. As a result, there has not been a resounding success in the didactic use of simulations, or at least not in any obvious way.

A large body of preliminary work towards such a standardization has been done in the USA by the Open Source Physics (OSP) program, which was supported among others by the National Science Foundation. Its goal was to create a thesaurus of partial solutions specifically for use in physical simulations, which could then be used in an object oriented way for programming specific simulations. The programs are made generally available under the GNU license-free open source model, with the obligation that the same applies to new third party solutions built on OSP.

GNU defines its goal as follows: "The GNU General Public License is intended to guarantee the freedom, to share and change all versions of a program. It aims to ensure that the software remains free for all its users. We, the Free Software Foundation, use the GNU General Public License for most of our software, it applies also to any other works, the authors of which have released in this way."

A leading pioneer of the OSP project was *Wolfgang Christian* at *Davidson College*, who built on a family of Java-**Physlets**, which he had developed previously. Together with his colleagues he created a number of program packages for the calculation and visualization of physics and engineering simulations, which included specific methods.

This was connected with the development of a curriculum for an introduction to the structure and technique of programming with *OSP*, and the development of a *launcher* by *Doug Brown*, which allows you to combine a whole sequence of simulations on similar topics as a course, inclusive of explanations, in a single file. This can be done quite compactly, since the simulations share a common set of data, which are only needed once in the launcher package. Individual simulations can be called from this file or can be isolated.

There is a wealth of partially simple and partially very refined physics simulations to be found in the OSP program and, in the following sections, we will briefly introduce the most important packages that are now available.

If one wants to fully understand a simulation file that has been created with OSP, one has to become quite familiar with its *html source code*. This is quite possible to do when using the teaching material, although it is still a difficult task. A second limitation of its general applicability lies in the fact that the visualization requires a great deal of effort and its development in html source code can be confusing for the less experienced user.

In this regard, the development of the **EJS package (Easy Java Programming)** by *Francisco Esquembre* was a further breakthrough for OSP. This package consists of a graphical user interface, which we briefly describe in the following. Its particular appeal is the possibility of taking the building blocks of the simulation from a large pre-built stock and to construct a realization tree from them via *drag and drop*. The individual icons are then connected to the simulation variables and to the easily selectable standard methods. For the creation of the proper calculation code, visual tools have also been provided. It is easy to become familiar with EJS using already existing examples, so that one does not need extensive technical knowledge of JAVA to develop simulations. Therefore EJS appears quite suitable for students of physics, whose main interest is in building physical models, rather than programming techniques.

Another big advantage of the EJS program in its current version is that, from inside the individual simulations, one can access the universal creation program, the **EJS console** via mouse click, so that one can immediately dive into the programming, make changes or take individual building blocks for one's new developments.

Thus the combination of **EJS + OSP** seems to be destined to become the standard program for didactically orientated simulations in the domain of physics and mathematics. We have used it almost exclusively in this work, although the author was previously more familiar with *Excel/VBA* and he first needed to become familiar with the new methods. The two links *EJS basic* and *EJS introduction* at the side margins lead to a description of EJS by Wolfgang Christian and Francisco Esquembre. In the same directory you will also find further documentation.

EJS and OSP are under active development, and thus are a *work in progress*: you are therefore advised to make yourself familiar with the current status using the supplied internet pages.

To use the Java simulations, the **Java Runtime Environment** must be installed at least in version *Java/re5*. You might want to install the free-of-charge current version (*June 2011: Java 6/update 24*) via the link on the margin.

11.2   A short introduction to EJS (Easy Java Simulation)

You reach the up-to-date description of the EJS program via the home page given in the link on the side. Here we give a very brief overview and suggest you have a close look at this program. When calling the EJS program on the lower boundary of the screen, the *EJS console* appears, as shown in Figure 11.1.

In the first line of the main window, one enters the directory where the **Java-JRE** (**Java runtime environemt**) is located, if the program does not find it by itself. In the second line, an arbitrary directory can be defined as **workspace** for EJS; the program then automatically creates two directories in this directory:

- **Source:** for *.xlm files;
- **Export:** for compressed *.jar files.

The program stores new or changed files automatically in these directories, unless other paths for saving are specified in individual cases. The two directories can contain a hierarchy of further directories. Files that have been automatically stored can later be copied or moved to other places.

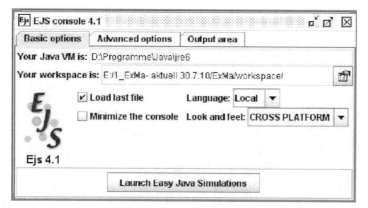

Figure 11.1. EJS console.

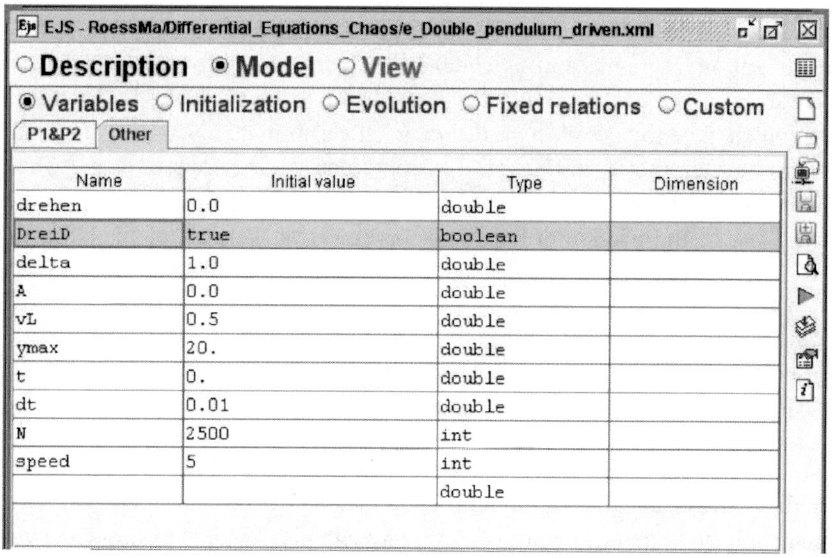

Name	Initial value	Type	Dimension
drehen	0.0	double	
DreiD	true	boolean	
delta	1.0	double	
A	0.0	double	
vL	0.5	double	
ymax	20.	double	
t	0.	double	
dt	0.01	double	
N	2500	int	
speed	5	int	
		double	

Figure 11.2. Simulation. The figure shows a page for the definition of parameters (variables). The simulation calls a working EJS console with editing window. It either contains no data, or those from the last simulation. You can browse through the individual pages and change entries. You may study the many possibilities in the main window **View**. Please be careful when saving in order not to overwrite any files; you should choose a name that does not already exist!

With **Launch Easy Java Simulation**, the editing window in Figure 11.2 is created. The console can be configured in such a way that this step happens automatically whenever it is called.

Figure 11.2 shows its visual interface. The main menu in the top line contains three sections, each of which can consist of several pages.

- **Description as text**;
- **Model** (Code);
- **View** (optical surface for the creation of the visualization tree).

For many simulations only a few pages are active. Simple function plotters, for example, do not require any special code and can be realized using *view* alone.

In the following line of the menu *model* in Figure 11.2 the sub-menu **Variables** is highlighted. The page **Other** for the simulaion **Double_pendulum_driven** is shown.

The individual pages of **Model** have the following meaning:

Variables: Global variables, which do not appear only locally in individual code methods. These are designated as decimal number (double), integers (int), symbolic text (string), logical variable (boolean) and are also classified according to their dimensions (for example number of computation loops with indices i, j, k: $[i]$ or $[i][j][k]\dots$). It is important to note that the decimal point appears as point, as is standard in the USA, and not as a comma as in German.

Initialization: Here the starting conditions are entered, for example specific values for variables, equations involving the variables or calls to methods listed on other pages, which have to take place at the start. This also includes logical equations that select from different possibilities. Help is provided via the context menu, which can be called with the right mouse button. The following example puts the two initial velocities v_a of the pendulum bobs and the time t at the start of the simulation to zero:

$$t = 0.0;$$

$$va1 = 0.0;$$

$$va2 = 0.0.$$

Evolution: Controls the succession of events, for example for an animation. It is particularly important that differential equations can be entered here, that are then automatically solved via a choice of different methods . Typical example:

$$\frac{dy_1}{dt} = va_1 \quad \text{to be solved with Runge–Kutta 4.}$$

Fixed Relations: Here relationships between the variables can be entered, which are always valid and provide input to the calculation. The following example from the simulation connects variables with trigonometric functions of other variables. (Please note that the function name must start with *Math.* in Java code).

$$x1 = L1*Math.sin(a1);$$

$$y1 = -L1*Math.cos(a1);$$

$$x2 = L2*Math.sin(a2);$$

$$y2 = -L2*Math.cos(a2).$$

Custom: Here special methods are formulated in *Java code*, which are called, for example, from the initialization page or by the control elements. The following example defines the action of the *clear* **button** in the view. It deletes all lines and resets the variables to their initial values.

public void clear(){_resetView();_initialize();}.

Here the methods _resetView() and _initialize() can be selected from a large number of prepared subroutines and one does not have to program these.

We now describe the functions of the icons on the right-hand side. The shorthand *. stands, as is usual for files, for an arbitrary filename before the file type (e.g. xml or jar).

Line 2, on the right: information about author and file;

Line 3 from the top: open new file;

Line 4 from the top: open *.xml-files;

Line 5 from the top: (*Opening*, with screen) leads to the homepage of EJS and a current library of EJS simulations;

Line 6 from the top: saving at the original location and under the same name as *.xml file, which can be opened from the console. *.xml files cannot be activated by themselves, but are very compact;

Line 7 from the top: saving the file at another location or under a different name as *.xml file;

Line 8 from the top: Searching help. After entering the keyword, it shows where it appears in the file;

Line 9 from the top (green triangle): creates the active simulation or gives an error message with hints;

Line 10 from the top: compressing the EJS file as *.jar file. Such files can be called as stand-alone applications and contain all required codes except for Java, which has to be installed on the computer. Alternatively html-pages or applets can be created;

Line 11 from the top: Opens general editing options, which are not required for creating files;

Line 12 from the top: Calling the internet help page of the EJS program.

View

Figure 11.3 shows a typical *View* page. The visualization tree is only partly visible.

On the right there are three menus below each other with a few pages each. They contain numerous icons that can be put together with *drag and drop* on the visualization tree.

The top menu, which is called **Interface**, includes *containers* as superior *parents* of the Java hierarchy, and pre-assembled control elements as *children* to be contained inside it.

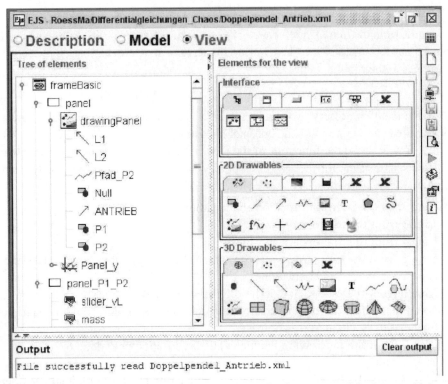

Figure 11.3. View page of EJS. On the right are the selection menus of the icons, which can be pulled with drag and drop into the visualization tree on the left.

The second menu **2D Drawables** contains icons that can be inserted for two-dimensional visualizations. The third menu **3D Drawables** contains icons for three-dimensional visualizations. In addition to icons that symbolize a single element, there are some that represent whole families of elements, for example arrows, points or curves.

Every icon used on the tree is showing after double clicking a large menu for formatting and for connecting to variables and methods. Figure 11.4 shows this for the relatively simple icon **P1**, which represents the bob of the main pendulum in the double pendulum simulation. The coordinates x_1 and y_1 are connected to the elements **Posx** and **Posy**. In the element **Size** the same dimensions are fixed in both directions. **Draggable** *True* means that the point can be pulled with the mouse, which automatically gives a new value to the variable. **On Drag** *Pendel()* calls the procedure *Pendel()* when the mouse is pulled (deleting previous traces and restarting the calculations). Many open positions can be used for further formatting.

The exact definition of individual elements appears if one holds the mouse pointer on the designation on the left. If one points at the label *Visible*, the message *The visibility of the element (boolean)* appears. If one double clicks on the first icon at the right of an element, either the existing choices (for visible *true, false* are shown,

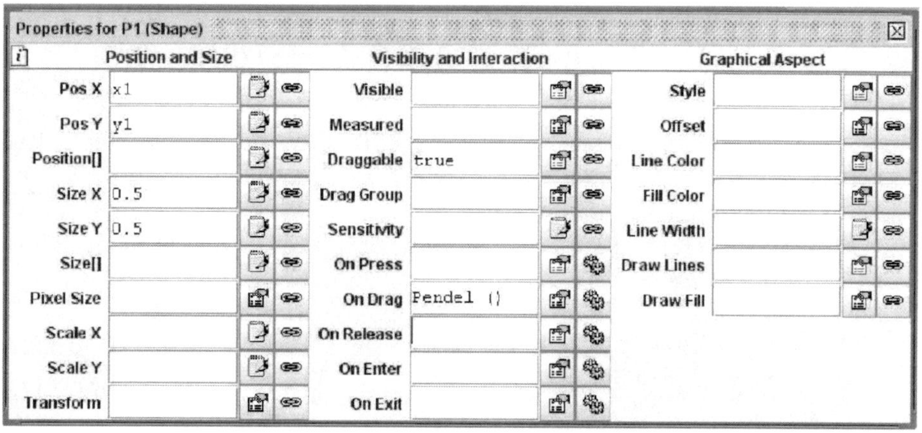

Figure 11.4. Window for fixing the properties of a visualization element (here of a point P1).

or assistance for entering information is given. The second icon contains a list of permitted quantities or methods, each of which can be selected with a click.

The next Figure 11.5 finally shows the appearance of the main window in the active simulation. You can certainly identify elements of the visualization tree, especially on the left-hand side (L1 and L2 are the pendulum bars, P1 and P2 are the pendulum bobs, Pfad_P2, the red orbit of the secondary pendulum, and so on).

You actually do not need to know any more to start with the simulations that have been developed for this book, and to use and change them. The same applies to other simulations that have been created with *EJS*. The **EJS console** required for this purpose is contained in the data carrier for this book in the version of February 2011. You can download a possibly newer version from the *EJS* homepage.

Start with something simple, for example with the calculation of the **geometric series**. Pressing the *ctrl* key and clicking on *Simulation* in the caption of Figure 11.6 opens the working EJS simulation as an independent **.jar file*.

Now click with the right mouse button on the simulation and choose **Open EJS Model** in the context menu that appears. A menu will pop up, which shows how an **.xml* file is extracted and stored. The standard storage location is the directory **source** in the **EJS workspace**; you may also choose a different one.

After confirming your input, the EJS console appears with the editing window as in Figure 11.3. The previously active simulation vanishes into the background, and a passive EJS window appears. You may see the configuration of the elements, but the simulation cannot run in these windows.

Now save under another name (the *.xml file is saved). Pressing the icon in the shape of a **green triangle** creates an active simulation under a new name.

Now change individual elements in the pages of the editing console and then save the *.xml file under a new name. If you close the old version and click on the green

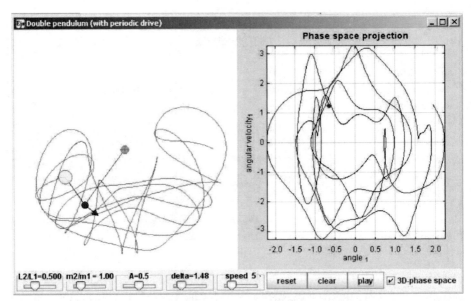

Figure 11.5. Simulation. Visualization window for the simulation of the double pendulum. The left side of the window shows the two masses (primary pendulum as small blue circle, secondary pendulum as large yellow circle) of the double pendulum, which is fixed and rotatable in the green point. The plane orbit of the yellow pendulum mass is drawn in red; it is very irregular. The right side of the window shows the phase-space of the primary pendulum.

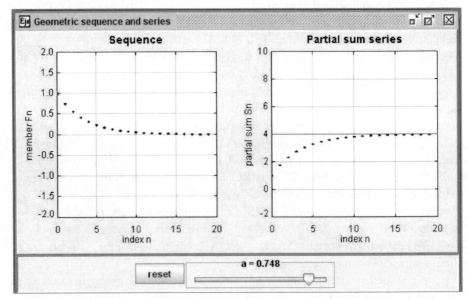

Figure 11.6. Simulation. Geometric sequence and series as a simple practice example. After starting, the EJS window is opened by clicking on the simulation window with the right mouse button.

triangle, your version becomes active, or you get an error message with solution hints if you have built a *bug* into the program. Initially, just go into *View* and change the colors or the thickness of lines. This way nothing can really break.

With some more insight you can also change the simple formulas in *Custom* and *Fixed Relations* and thus calculate a different series. Frequent saving under different names allows you to find working versions when encountering errors.

The *help* function, which can be called with the link on the margin or directly from `help/doc` the *console*, contains details on all individual elements. At this point you also find extensive documentation, for example for an introductory course.

11.3 Published EJS simulations

The following list consists of 100 EJS simulation that were available directly from the OSP home page in 2009. The list contains the title, link and the beginning of the description, as taken from the home page. After following links, further details can be found on the OSP homepage:

- complete description
- level and user
- keywords about the specialization area
- authors
- often a typical picture.

Following the links coupled to the names within the list takes you to **the download pages**, where you have two options:

Firstly: loading of the executable *.jar file of typically 1–2 MB. Secondly: Loading of the *.xml file of typically 10–100 kB, optionally with additional picture files, packaged into a *.zip file, which can be opened by the EJS console. The picture files contain elements of the description pages that do not belong to the standard thesaurus, such as formulas in special formats, pictures, drawings, and so on.

Files that have been opened with the console can be saved as compressed *.jar files, such that a duplication of the download is not necessary.

On the other hand, it is is possible for all *.jar files contained in the list to generate the *.xml file and the picture files from inside the active simulation using the context menu of the right mouse button, or to open the console.

The following list is roughly ordered according to subjects, to provide an easy overview. The numbers next to the titles correspond to their position on the OSP homepage.

You can quickly and directly call the *.jar files via the links that are shown next to the list at the margin. They immediately lead to a file that is already saved on your data carrier, and which is already executable and interactive.

11.3.1 Electrodynamics

28. Magnetic Field from Loops Model

The EJS Magnetic Field from Loops model computes the B-field created by an electric `*.jar`
current through a straight wire, a closed loop, and a solenoid.

62. Electromagnetic Wave Model

The EJS Electromagnetic Wave model displays the electric field and magnetic field `*.jar`
of an electromagnetic wave. The simulation allows an arbitrarily polarized wave to be
created.

11.3.2 Fields and potentials

9. Scalar Field Gradient Model

The Scalar Field Gradient Model displays the gradient of a scalar field using a numer- `*.jar`
ical approximation to the partial derivatives. This simple teaching model also shows
how to display and model scalar and ...

30. Lennard-Jones Potential Model

The EJS Lennard-Jones Potential model shows the dynamics of a particle of mass m `*.jar`
within this potential. You can drag the particle to change its position and you can drag
the energy-line to change its total energy. The ...

31. Molecular Dynamics Model

The EJS Molecular Dynamics model is constructed using the Lennard-Jones potential `*.jar`
truncated at a distance of three molecular diameters. The motion of the molecules is
governed by Newton's laws, approximated using ...

33. Molecular Dynamics Demonstration Model

The EJS Molecular Dynamics Demonstration model is constructed using the Lennard- `*.jar`
Jones potential truncated at a distance of three molecular diameters. The motion of the
molecules is governed by Newton's laws, approximated ...

11.3.3 Mathematics, differential equations

1. Linear Congruent Number Generator

The Linear Congruent Number Generator Model. The method generates a sequence `*.jar`
of integers x_i over the interval $0, m - 1$ by the recurrence relation $x_i + 1 = (ax_i + c)$
mod m where the modulus m is greater ...

3. Uniform Spherical Distribution Model

The EJS Uniform Spherical Distribution Model shows how to pick a random point `*.jar`
on the surface of a sphere. It shows a distribution generated by (incorrectly) picking
points using a uniform random distribution ...

6. Binomial Distribution Model

The EJS Binomial Distribution Model calculates the binomial distribution. You can ▀.jar ▀▀▀
change the number of trials and probability. You can modify this simulation if you
have EJS installed by right-clicking within ...

16. Great Circles Model

The EJS Great Circles model displays the frictionless motion of a particle that is ▀.jar ▀▀▀
constrained to follow the surface of a perfect sphere. The sphere rotates underneath
the particle, but since there is no ...

20. Cellular Automata Rules Model

The EJS Cellular Automata Rules Model shows a spatial lattice which can have any ▀.jar ▀▀▀
one of a finite number of states and which are updated synchronously in discrete time
steps according to a local (nearby neighbor) ...

21. Cellular Automata (Rule 90) Model

The EJS Cellular Automata (Rule 90) model displays a lattice with any one of a finite ▀.jar ▀▀▀
number of states which are updated synchronously in discrete time steps according to
a local (nearby neighbor) rule. Rule ...

24. Special Functions Model

The EJS Special Functions Model shows how to access special functions in the OSP ▀.jar ▀▀▀
numerics package. The simulation displays a graph of the special function over the
given range as well as the value of the selected ...

37. Harmonics and Fourier Series Model

The EJS Harmonics and Fourier Series model displays the sum of harmonics via a ▀.jar ▀▀▀
Fourier series to yield a new wave. The amplitude of each harmonic as well as the
phase of that harmonic can be changed via sliders ...

60. Fourier Sine Series

The Fourier sine series model displays the sine series expansion coefficients of an ▀.jar ▀▀▀
arbitrary function on the interval $[0, 2pi]$.

90. Poincare Model

The EJS Poincare model computes the solutions to the set of non-linear equations, ▀.jar ▀▀▀
$x' = x(a - b + z + d(1 - z^2)) - cy$, $y' = y(a - b + z + d(1 - z^2)) + cx$,
$z' = az - (x^2 ...$

91. Hénon-Heiles Poincare Model

The EJS Hénon-Heiles Poincare model computes the solutions to the non-linear ▀.jar ▀▀▀
Hénon-Heiles Hamiltonian, which reads $1/2(px^2 + py^2 + x^2 + y^2) +^2 y- ...$

92. Duffing Poincare Model

The EJS Duffing Poincare model computes the solutions to the non-linear Duffing `*.jar`
equation, which reads $x'' + 2\gamma x' - x(1 - x^2) = f\cos(\omega t)$, where each prime denotes
a time derivative. . . .

93. Duffing Phase Model

The EJS Duffing Phase model computes the solutions to the non-linear Duffing equa- `*.jar`
tion, which reads $x'' + 2\gamma x' - x(1 - x^2) = f\cos(\omega t)$, where each prime denotes a
time derivative. . . .

94. Duffing Measure Model

The EJS Duffing Measure model computes the solutions to the non-linear Duffing `*.jar`
equation, which reads $x'' + 2\gamma x' - x(1 - x^2) = f\cos(\omega t)$, where each prime denotes
a time derivative. . . .

95. Duffing Chaos Model

The EJS Duffing Chaos model computes the solutions to the non-linear Duffing equa- `*.jar`
tion, which reads $x'' + 2\gamma x' - x(1 - x^2) = f\cos(\omega t)$, where each prime denotes a
time derivative. . . .

96. Duffing Baker's Map Model

The EJS Duffing Baker's Map model computes the solutions to the non-linear Duffing `*.jar`
equation, which reads $x'' + 2\gamma x' - x(1 - x^2) = f\cos(\omega t)$, where each prime denotes
a time derivative. . . .

97. Duffing Attractor Model

The EJS Duffing Attractor model computes the solutions to the non-linear Duffing `*.jar`
equation, which reads $x'' + 2\gamma x' - x(1 - x^2) = f\cos(\omega t)$, where each prime denotes
a time derivative. . . .

98. Duffing Oscillator Model

The EJS Duffing Oscillator model computes the solutions to the non-linear Duffing `*.jar`
equation, which reads $x'' + 2\gamma x' - x(1 - x^2) = f\cos(\omega t)$, where each prime denotes
a time derivative. . . .

99. Baker's Map Model

The EJS Baker's Map model computes a class of generalized baker's maps defined `*.jar`
in the unit square. The simulation displays the resulting points as well as the distance
between adjacent points. The starting . . .

11.3.4 Mechanics

7. Mechanics Package: Challenging Intro Physics Topics

The EJS Mechanics Package: Challenging Intro Physics Topics contains Easy Java *.jar
Simulations (EJS) models used in a high-level Introductory Physics course for physics
majors. The topics include vector kinematics ...

11. Slipping and Rolling Wheel

The EJS Slipping and Rolling Wheel Model shows the motion of a wheel rolling on a *.jar
floor subject to a frictional force as determined by the coefficient of friction μ_k. The
simulation allows the user to change ...

23. Ceiling Bounce Model

The EJS Ceiling Bounce Model shows a ball launched by a spring-gun in a building *.jar
with a very high ceiling and a graph of the ball's position or velocity as a function of
time. Students are asked to set the ball's ...

25. Two Particle Elastic Collision Model

The EJS Elastic Collision Model allows the user to simulate a two-dimensional elastic *.jar
collision between hard disks. The user can modify the mass, position and velocity of
each disk using the sliders. Both ...

41. Baton Throw Model

The EJS Baton Throw model displays a baton thrown up in the air about its center of *.jar
mass. The baton is modeled by two masses separated by massless rigid rod. The path
of the center of mass of the baton and ...

42. Rocket Car on an Inclined Plane Model

The EJS Rocket Car on an Inclined Plane model displays a car on an inclined plane. *.jar
When the car reaches the bottom of the incline, it can be set to bounce (elastic collision) with the stop attached to the ...

43. Car on an Inclined Plane Model

The EJS Car on an Inclined Plane model displays a car on an incline plane. When the *.jar
car reaches the bottom of the incline, it can be set to bounce (elastic collision) with
the stop attached to the bottom ...

44. Kinematics of a Translating and Rotating Wheel Model

The EJS Kinematics of a Trainslating and Rotating Wheel model displays the model *.jar
of wheel rolling on a floor. By controlling three variables, the kinematics of the wheel
can be changed to present sliding, ...

46. Roller Coaster

The EJS Roller Coaster Model explores the relationship between kinetic, potential, *.jar
and total energy as a cart travels along a roller coaster. Users can create their own
roller coaster curve and observe the ...

47. Energizer

The EJS Energizer model explores the relationship between kinetic, potential, and ***.jar** total energy. Users create a potential energy curve and observe the resulting motion. The Energizer model was created using …

53. Inelastic Collision of Particles with Structure Model

The EJS Inelastic Collision of Particles with Structure model displays the inelastic ***.jar** collision between two equal "particles" with structure on a smooth horizontal surface. Each particle has two microscopic …

55. Platform on Two Rotating Cylinders Model

The EJS Platform on Two Rotating Cylinders model displays the model of a platform ***.jar** resting on two equal cylinders are rotating with opposite angular velocities. There is kinetic friction between each cylinder …

67. Two Falling Rods Model

The EJS Two Falling Rods model displays the dynamics of two rods which are dropped ***.jar** on a smooth table. In one case the end point on the table slides without friction, while in the other case it rotates about …

68. Coin Rolling without Sliding on an Accelerated Platform Model

The EJS Coin Rolling without Sliding on an Accelerated Platform model displays the ***.jar** dynamics of a coin rolling without slipping on an accelerated platform. The simulation dis-plays the motion of the coin as …

69. Coin Rolling with and without Sliding Model

The EJS Coin Rolling with and without Sliding model displays the dynamics of an ***.jar** initially rotating, but not translating, coin subject to friction. The simulation displays the motion of the coin as well as …

70. Orbiting Mass with Spring Force Model

The EJS Orbiting Mass with Spring Force model displays the frictionless dynamics ***.jar** of a mass constrained to orbit on a table due to a spring. The simulation displays the motion of the mass as well as the effective …

85. Symmetric Top Model

The EJS Symmetric Top model displays the motion of a top, in both the space frame ***.jar** and body frame, with no net toque applied. The top has an initial angular speed in the x, y, and z directions. The moments …

86. Lagrange Top Model

The EJS Lagrange Top model displays the motion of a heavy symmetric top under the ***.jar** effect of gravity. The top has an initial angular speed that provides the precessional, nutational, and rotational speeds …

87. Torque Free Top Model

The EJS Torque Free Top model displays the motion of a top, in both the space frame and body frame, with no net toque applied. The top has an initial angular speed in the x, y, and z directions. The moments ...

88. Falling Rod Model

The EJS Falling Rod model displays the dynamics of a falling rod which rotates about a pivot point as compared to a falling ball. The simulation allows computing fall times and trajectories. The initial ...

89. Spinning Dumbbell Model

The EJS Spinning Dumbbell model displays the motion of a dumbbell spinning around the fixed vertical axis z with constant angular velocity. The trajectories of each mass as well as the system's angular velocity, ...

11.3.5 Newton

8. Classical Helium Model

The EJS Classical Helium Model is an example of a three-body problem that is similar to the gravitational three-body problem of a heavy sun and two light planets. The important difference is that the helium ...

71. Two Orbiting Masses with Relative Motion Model

The EJS Two Orbiting Masses with Relative Motion model displays the dynamics of two masses orbiting each other subject to Newtonian gravity. The simulation displays the motion of the masses in the inertial ...

72. Orbiting Mass with Constant Force Model

The EJS Orbiting Mass with Constant Force model displays the dynamics of an orbiting mass due to a constant force (a linear potential energy function). The simulation displays the motion of the mass as well ...

100. Newtonian Scattering Model

The EJS Newtonian Scattering model displays the gravitational scattering of a multiple masses incident on a target mass. The simulation displays the motion of the smaller. The number of particles and their ...

11.3.6 Optics

2. Two-Color Multiple Slit Diffraction

The Two-Color Multiple Slit Diffraction Model allows users to explore multiple slit diffraction by manipulating characteristics of the aperture and incident light to observe the resulting intensity. An exploration ...

26. Multiple Slit Diffraction Model

The EJS Multiple Slit Diffraction model allows the user to simulate Fraunhofer diffraction through single or multiple slits. The user can modify the number of slits, the slit width, the slit separation and ...

`*.jar`

40. Thick Lens Model

The EJS Thick Lens model allows the user to simulate a lens (mirror) by adjusting the physical properties of a transparent (reflecting) object and observing the object's effect on a beam of light. The user ...

`*.jar`

48. Optical Resolution Model

The EJS Optical Resolution model computes the image from two point sources as seen through a circular aperture such as a telescope or a microscope. The simulation allows the user to vary the distance between ...

`*.jar`

64. Brewster's Angle Model

The EJS Brewster's Angle model displays the electric field of an electromagnetic wave incident on a change of index of refraction. The simulation allows an arbitrarily linearly (in parallel and perpendicular ...

`*.jar`

78. Interference with Synchronous Sources Model

The EJS Interference with Synchronous Sources model displays the interference pattern on a screen due to between one and twenty point sources. The simulation allows an arbitrarily superposition of the sources ...

`*.jar`

83. Two Source Interference Model

The EJS Two Source Interference model displays the interference pattern on a screen due to two point sources. The simulation allows an arbitrarily superposition of the two sources and shows both the current ...

`*.jar`

11.3.7 Oscillators and pendulums

15. Inertial Oscillation Model

The EJS Inertial Oscillation model displays the motion of a particle moving over the surface of an oblate spheroid. The spheroid is flattened to an ellipsoid of revolution because it is rotating, just as the ...

`*.jar`

17. Foucault Pendulum Model

The EJS Foucault Pendulum model displays the dynamics of a Foucault pendulum. The simulation is designed to show the dynamical explanation of why precession of the Foucault pendulum is slower at lower latitudes ...

`*.jar`

18. Circumnavigating Pendulum Model

The EJS Circumnavigating Pendulum model displays the dynamics of a mechanical oscillator in uniform circular motion. The mechanical oscillator is free to move in two directions. This 2-dimensional simulation . . .

34. Strange Harmonic Oscillator Model

The EJS Strange Harmonic Oscillator model displays the motion of two masses con- nected by a massless rigid rod, and the masses may move without friction along two perpendicular rails in a horizontal table . . .

35. Quartic Oscillator Model

The EJS Quartic Oscillator model displays the motion of a bead moving without fric- tion along a horizontal rod, while tied to two symmetric springs. Both the motion of the masses and the phase space plot are . . .

36. Damped Driven Harmonic Oscillator Phasor Model

The EJS Damped Driven Harmonic Oscillator Phasor model displays the motion of damped driven harmonic oscillator. The resulting differential equation can be extended into the complex plane, and the resulting . . .

38. Spring Pendulum Model

The EJS Spring Pendulum model displays the model of a hollow mass that moves along a rigid rod that is also connected to a spring. The mass, therefore, undergoes a combination of spring and pendulum oscillations . . .

39. Oscillator Chain Model

The EJS Oscillator Chain model shows a one-dimensional linear array of coupled harmonic oscillators with fixed ends. This model can be used to study the propagation of waves in a continuous medium and the . . .

45. Pendulum on an Accelerating Train Model

The EJS Pendulum on an Accelerating Train model displays the model of a pendulum on an accelerating train. The problem assumes that the pendulum rod is rigid and massless and of length $L = 2$, and the pendulum . . .

50. Coupled Oscillators and Normal Modes Model

The EJS Coupled Oscillators and Normal Modes model displays the motion of coupled oscillators, two masses connected by three springs. The initial position of the two masses, the spring constant of the three . . .

51. Spinning Hoop Model

The EJS Spinning Hoop model displays the model of a bead moving along a hoop which is spinning about its vertical diameter with constant angular velocity. Friction is negligible. The simulation displays . . .

56. Anisotropic Oscillator Model

The EJS Anisotropic Oscillator model displays the dynamics of a mass connected to `*.jar`
two opposing springs. The simulation displays the motion of the mass as well as the
trajectory plot. The initial position . . .

58. Oscillations and Lissajous Figures Model

The EJS Oscillations and Lissajous Figures model displays the motion of a superpo- `*.jar`
sition of two perpendicular harmonic oscillators. The simulation shows the result of
the superposition. The amplitude and . . .

73. Action for the Harmonic Oscillator Model

The EJS Action for the Harmonic Oscillator model displays the trajectory of a simple `*.jar`
harmonic oscillator by minimizing the classical action. The simulation displays the
endpoints of the motion (t, x) which . . .

11.3.8 Quantum mechanics

27. Circular Well Superposition Model

The Circular Well Superposition simulation displays the time evolution of the position- `*.jar`
space wave function in an infinite 2D circular well. The default configuration shows the
first excited state with zero . . .

49. QM Eigenstate Superposition Demo Model

The EJS QM Eigenstate Superposition Demo model displays the time dependence of `*.jar`
a variety of superpositions of energy eigenfunctions for the infinite square well and
harmonic oscillator potentials. One of . . .

54. Barrier Scattering model

The EJS Barrier Scattering model shows a quantum mechanical experiment in which `*.jar`
an incident wave (particle) traveling from the left is transmitted and reflected from a
potential step at $x = 0$. Although . . .

59. Free Particle Eigenstates

The free particle energy eigenstates model shows the time evolution of a superpos- `*.jar`
tion of free particle energy eigenstates. A table shows the energy, momentum, and
amplitude of each eigenstate.

61. Eigenstate Superposition

The fundamental building blocks of one-dimensional quantum mechanics are energy `*.jar`
eigenfunctions Psi(x) and energy eigenvalues E. The user enters the expansion coefficients into a table and the simulation . . .

74. Wave Packet Model

The EJS Wave Packet model displays the motion of an approximate wave packet. The simulation allows an arbitrarily wave packet to be created. The default dispersion relation, with the frequency equal to the ...

11.3.9 Theory of relativity

65. Einstein's Train and Tunnel Model

The EJS Einstein's Train and Tunnel model displays the famous thought experiment from special relativity where a train enters a tunnel as seen from two points of view. In one case the train is seen in the ...

66. Simultaneity Model

The EJS Simultaneity model displays the effect of relative motion on the relative ordering of the detection of events. The wave source and two equidistant detectors are at rest in reference frame S', which ...

11.3.10 Statistics

4. Random Walk 2D Model

The EJS Random Walk 2D Model simulates a 2-D random walk. You can change the number of walkers and the probability of going a given direction. You can modify this simulation if you have EJS installed by right-clicking ...

5. Random Walk 1D Continuous Model

The EJS Random Walk 1D Continuous Model simulates a 1-D random walk with a variable step size. You can change the number of walkers and the probability of going right and left. You can modify this simulation if ...

29. Balls in a Box Model

The Balls in a Box model shows that a system of particles is very sensitive to its initial conditions. In general, an isolated system of many particles that is prepared in a nonrandom configuration will change ...

32. Multiple Coin Toss Model

The EJS Multiple Coin Toss model displays the result of the flipping of N coins. The result of each set of coin flips is shown by the image of the pennies on the screen and the complete results of the tossing ...

11.3.11 Thermodynamics

10. Kac Model

The EJS Kac Model simulates the relaxation of a gas to equilibrium by randomly `*.jar` selecting and then colliding gas molecules but without keeping track of the molecules' positions. As long as the collisions are ...

12. 2D-Ising Model

The EJS 2D-Ising model displays a lattice of spins. You can change the lattice size, `*.jar` temperature, and external magnetic field. You can modify this simulation if you have EJS installed by right-clicking within ...

11.3.12 Waves

52. Beats Model

The EJS Beats model displays the result of adding two waves with different frequen- `*.jar` cies. The simulation displays the superposition of the two waves as well as a phasor diagram that shows how the waves add ...

57. Normal Modes on a Loaded String Model

The EJS Normal Modes on a Loaded String model displays the motion of a light `*.jar` string under tension between two fixed points. The string is also loaded with N masses located at regular intervals. The number ...

63. Doppler Effect Model

The EJS Doppler Effect model displays the detection of sound waves from a moving `*.jar` source and the change in frequency of the detected wave via the Doppler effect. In addition to the wave fronts from the source ...

75. Waveguide Model

The EJS Waveguide model displays the motion of a traveling wave forced to move `*.jar` between two walls in a waveguide. The two walls are located at $y = 0$ and a, so that its normal modes are $u(t, x) = A \sin(n\pi$ ...

76. Waves and Phasors Model

The EJS Waves and Phasors model displays the motion of a transverse wave on a `*.jar` string and the resulting phasors for the wave amplitude. The simulation allows an arbitrarily polarized wave to be created. The ...

77. Transverse Wave Model

The EJS Transverse Wave model displays the motion of a transverse wave on a string. `*.jar` The simulation allows an arbitrarily polarized wave to be created. The magnitude of the components of the wave and the ...

79. Reflection and Refraction between Taut Strings Model

The EJS Reflection and Refraction between Taut Strings model displays the motion `*.jar` of a traveling pulse on a string when it is incident on a change of string density ...

80. Standing Waves on a String Model

The EJS Standing Waves on a String model displays the motion of a standing wave `*.jar` on a string. The standing wave can be augmented by adding the zero line and the maximum displacement of the string. The number ...

81. Resonance in a Driven String Model

The EJS Resonance in a Driven String model displays the displacement of taut string `*.jar` with its right end fixed while the left end is driven sinusoidally. The driving frequency, amplitude, and the simulation's ...

82. Standing Waves in a Pipe Model

The EJS Standing Waves in a Pipe model displays the displacement and pressure `*.jar` waves for a standing wave in a pipe. The pipe can be closed on both ends, on one end, or open on both ends. The number of nodes ...

84. Group Velocity Model

The EJS Group Velocity model displays the time evolution for the superposition of `*.jar` two traveling waves of similar wave numbers and frequencies. The simulation allows an arbitrarily superposition of two waves ...

11.3.13   Miscellaneous

13. Radioactive Decay Events Model

The EJS Radioactive Decay Events Model simulates the decay of a radioactive sample `*.jar` using discrete random events. It displays the number of events (radioactive decays) as a function of time in a given time ...

14. Radioactive Decay Distribution Model

The EJS Radioactive Decay Distribution Model simulates the decay of a radioactive `*.jar` sample using discrete random events. It displays the distribution of the number of events (radioactive decays) in a fixed time ...

19. Game of Life Model

The EJS Game of Life Model simulates a popular 2D cellular automaton of a lattice `*.jar` in a finite state which is updated in accordance with a set of nearby-neighbor rules. The universe of the Game of Life, developed ...

22. Radioactive Decay Model

The EJS Radioactive Decay Model simulates the decay of a radioactive sample using `*.jar` discrete random events. It displays the number of radioactive nuclei as a function of time. You can change the initial number ...

A large number of older EJS examples, among them very elementary ones, are found in the **users** directory that belongs to the directory tree of this book. They can also be downloaded together with the EJS console from the EJS home page. In the users directory the files are ordered by author. There are .xlm files in this directory, which are not executable by themselves and have to be loaded by the EJS console. The following Figure 11.7 of the directory tree will facilitate the orientation.

The directories of the authors are located below the directory **source/users**. In Figure 11.7 the directory tree details the sub-directory of *Francisco Esquembre*: *Murcia/ Fem* (University of Murcia, Spain).

Figure 11.7. *e-ExMath* is the root directory of this work with the corresponding textfile and the EJS console. In *doc* you will find the program descriptions of EJS. In *workspace* the executable *.jar files are in the directory *export*, while the directory *source* contains the *.xml files that have to be loaded from the EJS console. *Other* contains simulations from different sources: for the University of *Murcia* (Esquembre) the directory tree is recursed down to the actual simulation.

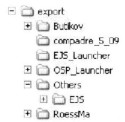

Figure 11.8. The directory *Export/Others/EHS* contains directly executable *.jar files.

To enable the user to get an overview of the large number of simulations available, the directory *export* contains in its sub-directory *Others/EJS* directly runnable *.jar files of the simulations next to the corresponding *.xml files (see Figure 11.8).

Using the hyperlink on the margin next to Figure 11.9 one reaches an overview file, which contains information on 144 simulations ordered according to 16 topics, supplemented with short comments and a reference for the respective source. The individual simulations can be called quickly and directly via clicking on the file name.

These files are of very different levels of complexity. In addition to a few child-friendly simulations, there are simple examples for the demonstration of certain visualization possibilities. The majority of the files contains rather complex simulations of physical problems, with optical visualizations that are, in a number of cases, quite convincing. Some of the simulations can also be found under the new individual files, which were discussed at the beginning of the chapter. Some of these have been developed further.

Many of the files contain no description pages. Testing which elements of the graphics can be pulled with the mouse often reveals initially unexpected design possibilities.

The files can be edited and further developed, if the corresponding *.xml file is called from the EJS console.

List of selected simulations from the EJS package (*.jar- files)   Dieter Roess

The corresponding *xml* - files have the same name, with capital first letter

The 144 files are sorted into 16 thematic groups

Close the running simulation before opening another one.

Theme	Remarks	Hyperlink to jar- file	Authors	
1 - Mechanics	Torque, point of action	at_moment	VonSiebenthal	Table
1 - Mechanics	Regular and chaotic oscillations	ball in Wedge	ejs_tpt_modeling	
1 - Mechanics	2D Collision, variable	collision2D_e	FuKwuHwang	
1 - Mechanics	Gyro with gravitation	lagrange	ehu_jma	
1 - Mechanics	Collision, also with repulsion!	multiple Collisions	ejs_crcExamples	
1 - Mechanics	Newton cradle, variable	newtonsCraddle	murcia_fem	

Figure 11.9. Beginning of the overview table.

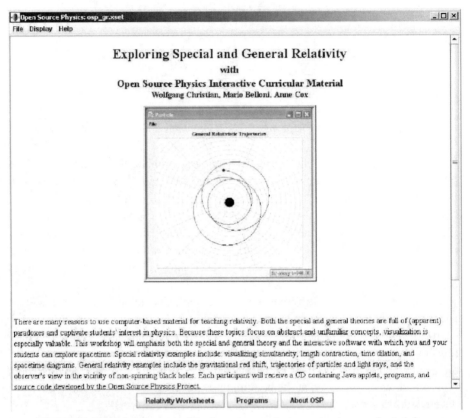

Figure 11.10. Opening page of the launcher package on the theory of relativity.

11.4 OSP Simulations that were not created with EJS

A large number of OSP Java simulations are present in a group of *Launcher* packages, structured according to themes, which can be obtained at the OSP homepage.

Many of the packages have been developed as courses. Figure 11.10 shows the typical appearance when opening one of them. This launcher has three directories that can be opened by clicking on the buttons or with the file menu.

The directory *Relativity Workshop*, which can be called from the list at the bottom, contains a complete course on special and general theory of relativity, subdivided into chapters ordered according to topics. Some of them contain descriptive text with static pictures, theory and problems, many contain, in addition, interactive simulations.

The directory *About OSP* contains details about the authors, about the launcher method, with which many individual files can be combined into a package, and about options for the presentation, among them language selection, provided this was enabled by the authors of the simulations.

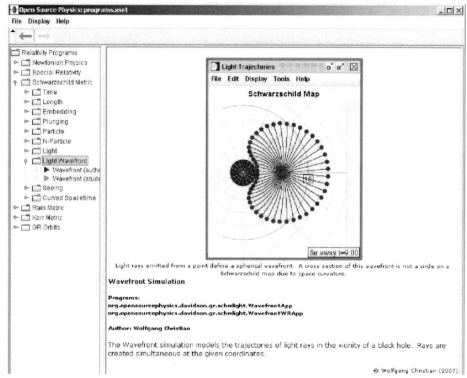

Figure 11.11. Directory structure of the page *Programs* of a typical *Launcher*. On the left a directory tree with numerous simulation files is shown. On the right a picture and a short description is shown for a selected simulation. The picture shows light rays that originate radially from a body in the vicinity of a black hole. Double clicking on the green triangle activates the simulation.

The directory *Programs* contains many interactive cosmological simulations from Newtonian mechanics to Kerr and Rain metrics. For a number of simulations, the version created by the lecturer are next to versions that were produced by students.

The directory *Programs* has the structure as shown in Figure 11.11, with many sub-directories.

The **File** menu on top of Figure 11.11 contains options for editing and for exporting individual simulations.

11.4.1   List of OSP launcher packages

In the following list we again show a table of the titles linked to the OSP homepage and the author's homepage, as well as a short description. In the margin is a link to directly access the launcher package on the data carrier.

1. Symmetry Breaking on a Rotating Hoop

The Rotating Hoop Launcher package shows the dynamics of a mass that is con- `Launc` strained to move on a rotating hoop. The rotating hoop model is an excellent mechanical model of first- and second-order phase transitions ...   Wolfgang Christian

2. Modeling a Changing World

Modeling a Changing World written by mathematics professor Tim Chartier and his `Launc` student Nick Dovidio presents curricular material in an OSP Launcher package to motivate the need for numerically solving ordinary ...   Tim Chartier

3. Hasbun Classical Mechanics Package

The Hasbun Classical Mechanics Package is a self-contained Java package of OSP `Launc` programs in support of the textbook "Classical Mechanics with MATLAB Applications". Classical Mechanics with MATLAB Applications ...   Javier Hasbun

4. Tracker Demo Package

The Tracker Sampler Package contains several video analysis experiments from me- `Launc` chanics and spectroscopy. It is distributed as a ready-to-run (compiled) Java archive containing the Tracker video analysis application, ...   Douglas Brown

5. Tracker Air Resistance Model

The Tracker Air Resistance Model asks students to explore air resistance of falling `Launc` coffee cups by considering both viscous (linear) and drag (quadratic) models. Students see a video of falling cups and explore ...   Douglas Brown

6. General Relativity (GR) Package

The General Relativity (GR) Package is a self-contained file for the teaching of gen- `Launc` eral relativity. The file contains ready-to-run OSP programs and a set of curricular materials. You can choose from a variety ...

Wolfgang Christian, Mario Belloni, Anne Cox

7. OSP QuILT Package

The OSP QuILT package is a self-contained file for the teaching of time evolution of `Launc` wave functions in quantum mechanics. The file contains ready-to-run OSP programs and a set of curricular materials.

Chandralekha Singh, Mario Belloni, Wolfgang Christian

8. Phase Matters Package

The Phase Matters package is a self-contained file for the teaching of phase and time `Launc` evolution in quantum mechanics. The file contains ready-to-run OSP programs and a set of curricular materials. The material ...   Mario Belloni, Wolfgang Christian

9. Spins Package

The Spins package is a self-contained file for the teaching of measurement and time `Launc` evolution of spin-1/2 systems in quantum mechanics. The file contains ready-to-run OSP programs and a set of curricular ...   Mario Belloni, Wolfgang Christian

10. Statistical and Thermal Physics (STP) Application

The Statistical and Thermal Physics (STP) Application is a self-contained file for `Launch`
the teaching of statistical and thermal physics. The file contains ready-to-run OSP
programs and a set of curricular materials. . . .   Harvey Gould, Jan Tobochnik

11. Momentum Space Package

The Momentum Space package is a self-contained file for the teaching of the time evo- `Launch`
lution and visualization of energy eigenstates and their superpositions via momentum
space in quantum mechanics. The file . . .   Mario Belloni, Wolfgang Christian

12. Position Carpet Package

The Position Carpet package is a self-contained file for the teaching of the time evo- `Launch`
lution and visualization of energy eigenstates and their superpositions via quantum
space-time diagrams or quantum carpets . . .   Mario Belloni, Wolfgang Christian

13. Wigner Package

The Wigner package is a self-contained file for the teaching of the time evolution and `Launch`
visualization of energy eigenstates and their superpositions in quantum mechanics.
The file contains ready-to-run OSP . . .   Mario Belloni, Wolfgang Christian

14. Modeling Physics with Easy Java Simulations: TPT Package

This Java archive contains a collection of simple Easy Java Simulations (EJS) programs `Launch`
for the teaching of computer-based modeling. The materials and text of this resource
appeared in an article of the same . . .   Wolfgang Christian, Francisco Esquembre

15. Superposition Package

The Superposition package is a self-contained file for the teaching of the time evo- `Launch`
lution and visualization of energy eigenstates and their superpositions in quantum
mechanics. The file contains ready-to-run . . .   Mario Belloni, Wolfgang Christian

16. Demo Package

The Demo package is a self-contained file for the teaching of orbits, electromagnetic `Launch`
radiation from charged particles and quantum mechanical bound states. The file contains ready-to-run OSP programs and . . .   Mario Belloni, Wolfgang Christian

17. Computer Simulation Methods Examples

Ready to run Launcher package containing examples for an Introduction to Computer `Launch`
Simulation Methods by   Harvey Gould, Jan Tobochnik, and Wolfgang Christian

18. OSP User's Guide Examples

Ready to run Launcher package containing examples for Open Source Physics: A `Launch`
User's Guide with Examples by   Wolfgang Christian

19. Numerical Time Development in Quantum Mechanics Using a Reduced Hilbert Space Approach

This self-contained file contains Open Source Physics programs for the teaching of [Launc] time evolution and visualization of quantum-mechanical bound states. The suite of programs is based on the ability to expand ...   Mario Belloni, Wolfgang Christian

Here you should again look directly at the OSP homepage, where new packages or versions are present.

In addition, the comfortable search function of the home page allows you to search for certain topics, levels and intended audiences. Figure 11.12 shows the search tree. The individual selection boxes are each structured into numerous categories. For the topics this is shown in Figures 11.12 and 11.13.

On the OSP homepage you also may find isolated simulations (429, as of November 2010). To find these you choose the search function under OSP type *Java Model*, as shown in Figure 11.14

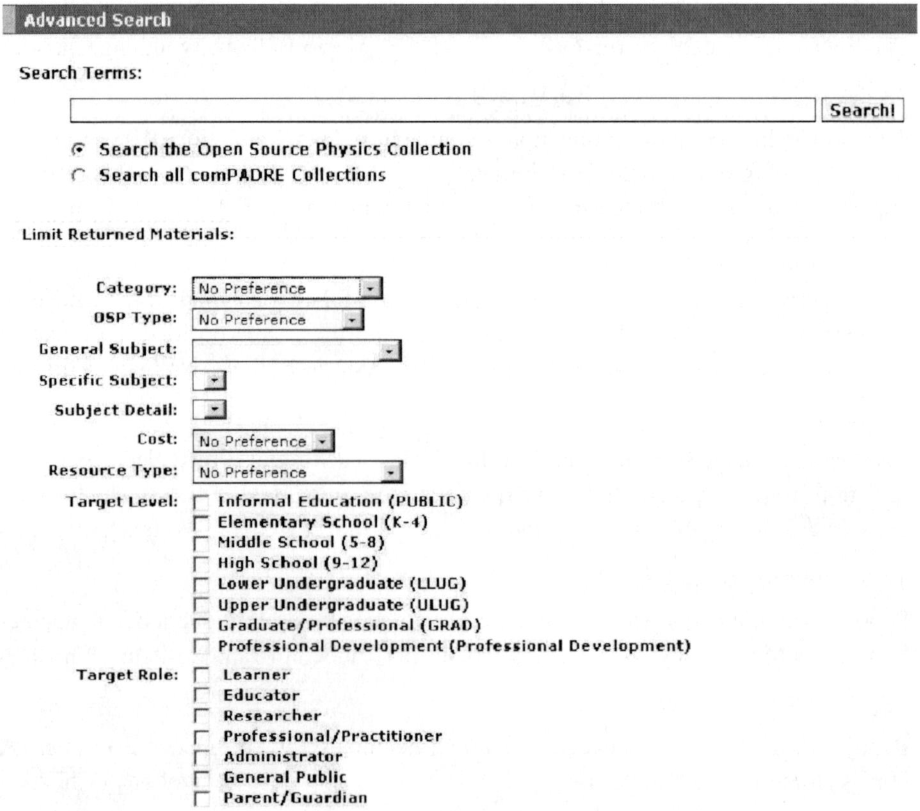

Advanced Search

Search Terms:

[   ] [Search!]

- Search the Open Source Physics Collection
- Search all comPADRE Collections

Limit Returned Materials:

Category: [No Preference ▾]
OSP Type: [No Preference ▾]
General Subject: [▾]
Specific Subject: [▾]
Subject Detail: [▾]
Cost: [No Preference ▾]
Resource Type: [No Preference ▾]

Target Level:
- Informal Education (PUBLIC)
- Elementary School (K-4)
- Middle School (5-8)
- High School (9-12)
- Lower Undergraduate (LLUG)
- Upper Undergraduate (ULUG)
- Graduate/Professional (GRAD)
- Professional Development (Professional Development)

Target Role:
- Learner
- Educator
- Researcher
- Professional/Practitioner
- Administrator
- General Public
- Parent/Guardian

[Searc]

Figure 11.12. Search window on the Compadre homepage.

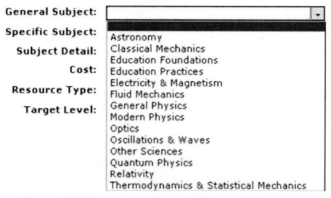

Figure 11.13. Topic selection on the Compadre homepage.

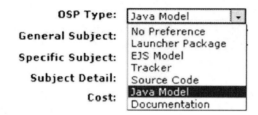

Figure 11.14. Method selection on the Compadre homepage.

Credit: We thank Wolfgang Christian and Francisco Esquembre for the permission to use the files as available on the *OSP* homepage (2010) in this book.

11.5   EJS simulations packaged as launchers

A number of physics courses, which demonstrate the application of the EJS program for the simulation of elementary and advanced physical problems, are also combined as *Launcher* packages. Individual solutions contained in them can be called via double click. The packages that are briefly described in the following are by Wolfgang Christian, Francisco Esquembre and their colleagues. They can be called directly via the link on the margin.

Ehu_mechanics-waves

course in mechanics, oscillations and waves   Juan M. Aguirregabiria   Launch

Ejs_crcExamples

description of EJS, many examples   Francisco Esquembre, Wolfgang Christian   Launch

Ejs_demo

description of EJS, simple examples from mechanics and thermodynamics, 3D visu-   `Launc`
alizations   Francisco Esquembre

Ejs_mabelloni_pendula

Lagrangian Mechanics, simple and complicated pendulums   Mario Mabelloni `Launc`

Ejs_mechanics

Basic course in mechanics and gas dynamics, with hints about the modeling technique `Launc`
                              Wolfgang Christian, Francisco Esquembre

Ejs_stp

Statistics and thermodynamics, FPU problem   Wolfgang Christian `Launc`

Ejs__tpt_modeling

Introduction to EJS and launcher packages, simple and advanced models from me- `Launc`
chanics and heat   Wolfgang Christian, Francisco Esquembre

Ejs_wochristian_chaos

Complex roots, Mandelbrot set, driven pendulum, phase space   `Launc`
                                                        Wolfgang Christian

Ejs_wochristian_examples

Advanced models, Fourier analysis, Lennard-Jones potential, oscillator chains `Launc`
                                                        Wolfgang Christian

Ejs_wochristian_odeflow

Some solutions of ordinary differential equations   Wolfgang Christian `Launc`

The advantage of these *EJS* launcher packages in comparison to the *OSP* packages discussed above, is that changing the simulations does not require advanced *JAVA* knowledge.

An active individual simulation can be transferred into the EJS console via the context menu (callable by clicking on the simulation with the right mouse button). In its windows, code and visualization elements can be seen and edited. Thus an existing solution can be quite easily used as starting point for further developments.

11.6 Cosmological simulations by *Eugene Butikov*

Because of their operating system independence we have, so far, only used Java simulations or given links to them.

Eugene Butikov (University of Petersburg) has created a large number of simulations for cosmological and other physical problems based on *Windows* and *Visual*

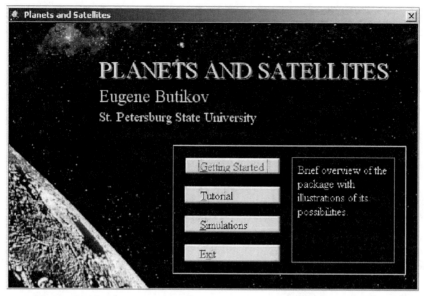

Figure 11.15. Simulation. Opening window of the *Butikov* simulations.

Basic. Their design is so convincing that we want to add them to our overview, although they will only be accessible to users who work with *Windows*. The Butikov program must be installed once for the simulations to run on your computer. You start the installation with the link *B. install* at the margin. After installation you can activate | B. install | the simulations via *Simulation* in the figure captions.

Because of the numerous possibilities, we will facilitate user access with a brief description. Figure 11.15 shows the start page of the program *Planets and Satellites*, whose content far exceeds what the title promises. Thus, in addition to elementary problems (Kepler's laws), it also treats many-body problems with their nonlinear and complex orbits, for example the passage of two stars with planets under *Planet robbery*. The graphical presentations are didactically very versatile. They show, for example, the time development, from the perspective of the star, of the planet or of the center of mass of the system (menu *View*), and at the same time yielding sometimes surprising orbits. The individual simulations (menu *Examples*) allow for many adjustments of all important parameters, so that the user can experiment freely.

The three parts of the package contain:

Getting started: Extensive hints for orientation; glossary of technical terms; and links to particularly appealing examples from the multitude of simulations.

Tutorial: Glossary; short overview of the course text; an extensive course text (accessible via the Menu *Help topics/Content*) and a linked table of contents, which leads directly to the individual simulations; didactic questions; help for handling the simulation.

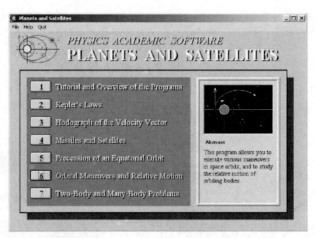

Figure 11.16. A selection window for characteristic groups of *Butikov* simulations. After choosing the group on the left-hand side, a typical picture with description appears on the right.

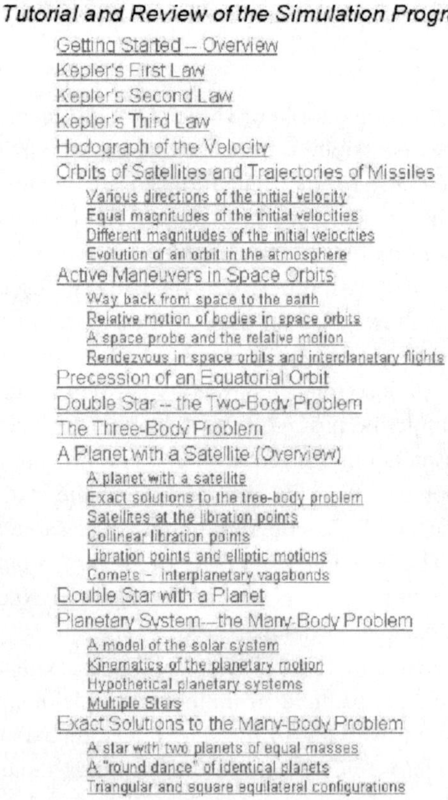

Tutorial and Review of the Simulation Programs

Getting Started – Overview
Kepler's First Law
Kepler's Second Law
Kepler's Third Law
Hodograph of the Velocity
Orbits of Satellites and Trajectories of Missiles
  Various directions of the initial velocity
  Equal magnitudes of the initial velocities
  Different magnitudes of the initial velocities
  Evolution of an orbit in the atmosphere
Active Maneuvers in Space Orbits
  Way back from space to the earth
  Relative motion of bodies in space orbits
  A space probe and the relative motion
  Rendezvous in space orbits and interplanetary flights
Precession of an Equatorial Orbit
Double Star – the Two-Body Problem
The Three-Body Problem
A Planet with a Satellite (Overview)
  A planet with a satellite
  Exact solutions to the tree-body problem
  Satellites at the libration points
  Collinear libration points
  Libration points and elliptic motions
  Comets – interplanetary vagabonds
Double Star with a Planet
Planetary System—the Many-Body Problem
  A model of the solar system
  Kinematics of the planetary motion
  Hypothetical planetary systems
  Multiple Stars
Exact Solutions to the Many-Body Problem
  A star with two planets of equal masses
  A "round dance" of identical planets
  Triangular and square equilateral configurations

Figure 11.17. Topics of individual simulations in the *Butikov* program and their descriptions.

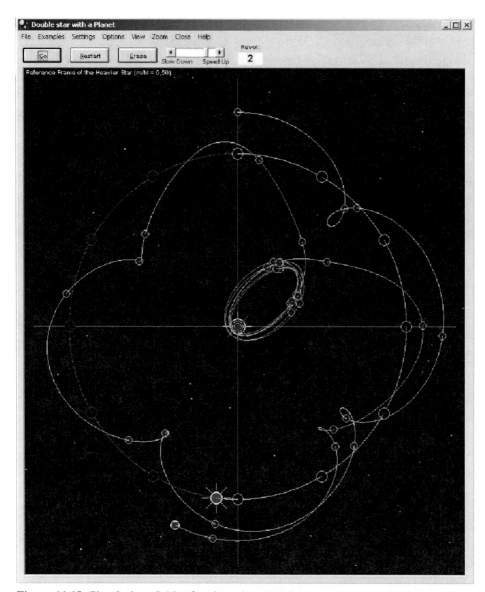

Figure 11.18. Simulation. Orbit of a planet in a double star system, seen in the coordinate system of the star with double the mass (red in the center). The lighter star travels around it on a close to circular orbit. The planet, which is very light in comparison with both stars, starts its yellow orbit on top around the blue star. This orbit is perturbed by the red central star and then moves over to a green orbit around it. After a few turns the perturbation by the blue star is sufficient to temporarily bind it to the blue star again (blue orbit).

Simulations: Access to the individual simulations, structured into seven classes, as shown in Figure 11.16.

Figure 11.17 (a screen-shot from *Tutorial*) shows the structure of the total program.

The link on the margin **Butikov** provides access to Butikov's homepage, from `Butikc` where you can see the programs he has published. There you will also find Java applets for many physics problems. The link **PAS** leads to the homepage of **Physics** `PAS` **Academics Software** (PAS), where the simulations were originally published.

With the permission of the author *Eugene Butikov* and the PAS editor *Jon Risley*, our collection contains the cosmological simulation program. You may call it with the interactive Figure 11.18.

It shows as an example a system of two stars of unequal mass with a common planet, whose orbits move from the one to the other star. This is displayed as seen from the coordinate system of the more massive star. The calculation of the orbit around the smaller star in yellow starts on top; the color changes to green when the planet moves into a orbit around the inner main star. Later, the planet again moves to the secondary star (blue section) of the orbit.

12 Conclusion

The development of this book has given me deeper insight into some foundations of mathematics, and has also given me great intellectual pleasure when experimenting with the didactic possibilities of the simulations. I wish the reader may benefit in his or her own striving for knowledge and be provided with a similar sense of achievement.